Questions, Questions, so many Questions

When was it built?

Why was it built?

How big is it?

Is this really the fourth site for the town of Dillon?

Where were the other sites?

What happened to the people and buildings in the old town?

Are there buildings under the water?

Some I could answer. Some I could not.

This book is the result of the search for answers.

Table of Contents

Dillon, Denver and the Dam 1

Early Diversion Plans 5

Geology of the Reservoir Site 9

The High Dam 17

The Roberts Tunnel 35

Geology along the Roberts Tunnel 37

Building the Roberts Tunnel 45

Old Dillon 47

New Dillon 73

Dillon Cemetery 83

Swan Mountain Road 87

Completing the Dam and Water Storage 89

Some Final Statistics 93

Time Line 97

How Dillon Received Its Name 99

New Dillon Walking Tour 103

References 111

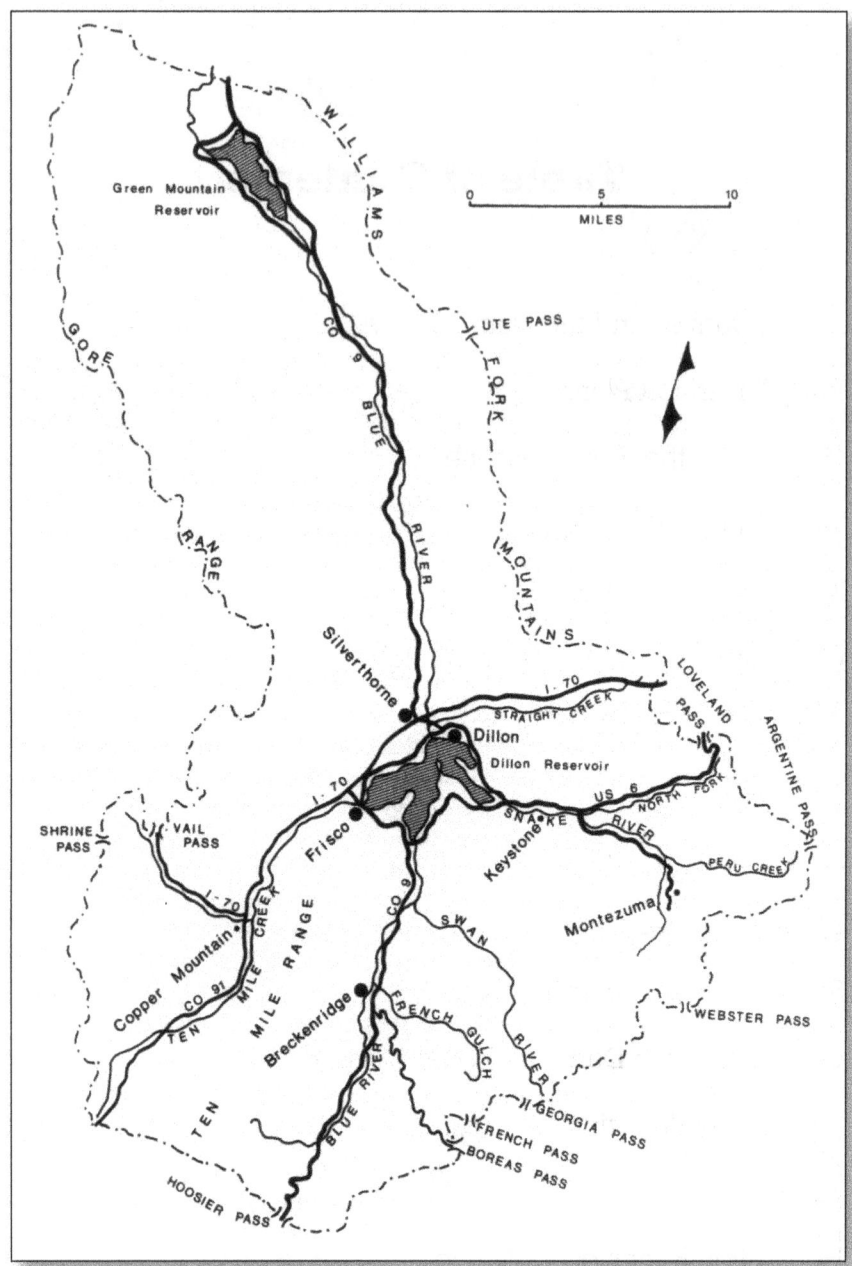

Figure 1. Summit County, Colorado. (Map drawn by Author)

Dillon, Denver and the Dam

THE BEAUTY OF LAKE DILLON brings smiles to the faces of many; but to others it brings back painful memories. To some, it symbolizes newness, growth, and development; to others, it recalls the death of a familiar, comfortable, and slower way of life. But, no matter how one feels about it or how one describes it, barring a major disaster, the dam and lake are here to stay. With it have come changes to a town's raison d'être, as well as in land use, recreation, scenic beauty, vegetation, stream flow, and to wildlife and the county's economy. *(Figures 3, 4)*

When Bayard Taylor passed through what is now Summit County in the summer of 1866, he camped near the present site of Dillon. The group's exact location is known because journals mention the confluence of three streams—the Blue, the Snake, and a third, unnamed one (the Ten Mile). Taylor's book, *Colorado: A Summer Trip,* tells of dense pines, thick green grass, hills of aspen, sagebrush in higher, drier areas, and fields of wildflowers. Four colors struck him: the gray of the sagebrush, the green of the meadows, and the brilliant white of the snow-covered peaks under a clear blue sky. They forded the "saddle-deep" and "swift" Snake with great diffi-

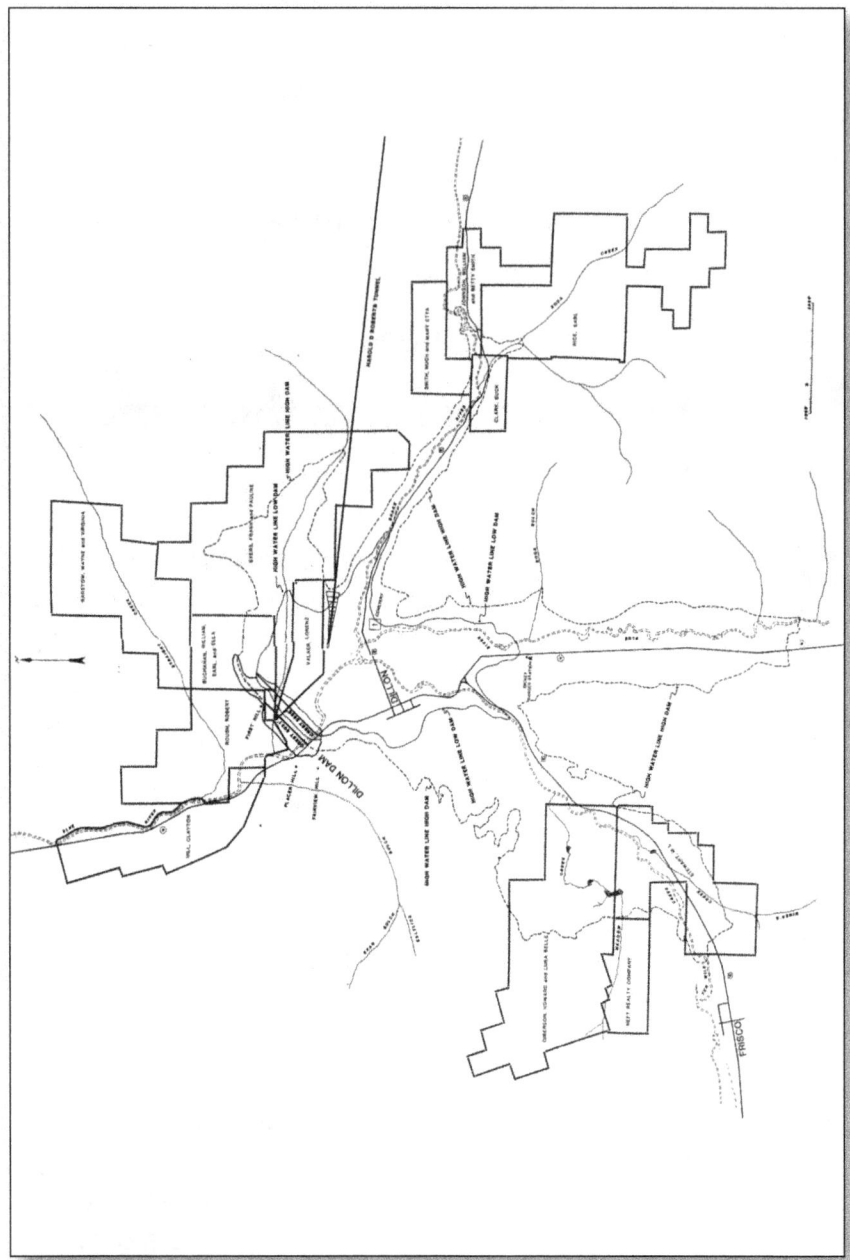

Figure 2. Map of Reservoir Site showing Selected Land Ownership, High and Low Dam Waterlines, and the Harold D. Roberts Tunnel. (Adapted from map provided by Board of Water Commissioners, City and County of Denver)

Figure 3. Dillon before the Dam—looking East. (Rocky Mountain View Company postcard. Courtesy Sena Valaer)

Figure 4. Same View following Completion of Dillon Dam—looking East. (Photograph by Author)

Figure 5. *Lake Dillon. The lake brings smiles to the faces of some but to others it brings back painful memories. (Photograph by Author)*

culty. The Blue, a "foaming" stream with an "impetuous" current, almost carried one horse and rider to their deaths. The raging water completely submerged one pack mule while it crossed the stream. The valley of the Blue, here narrowed by mountains, contained pools, quagmires, fallen timber, and rock slides.

To make the same trip today, Bayard Taylor would need a submarine. Gone are all of the landmarks he noted—buried under tons of water. He could no longer find his bearings using the confluence of the three streams. Nor would he find his swiftly flowing Snake, with "saddle-deep" water or his "foaming" Blue with "impetuous" current. ***(Figure 5)***

Early Diversion Plans

JUST AFTER THE TURN OF THE CENTURY, Denver, anticipating a rapid population growth, began developing a plan for using western slope water to quench the thirst of that eastern slope city.

W.H. Meyers and George J. Bancroft on October 29, 1907, proposed what would have been the first trans-mountain tunnel diversion project, the Collier Mountain Tunnel Project. This incomplete preliminary proposal would have taken Snake River water through the mountains near Montezuma. W.E. Goldsborough and Bancroft refiled the plan on January 13, 1909. A third filing, on April 10, 1910, called for a 4.2-mile-long Inter-Ocean tunnel that would take water from the Swan River to Jefferson Creek. The west portal of the tunnel would have an elevation of 10,322 feet. Ditches with elevations over 10,300 feet would bring water from the Ten Mile Creek, the Blue River and its tributaries, and the Snake River and its tributaries to the Swan River. One ditch, the Montezuma ditch, was to be 30.8 miles long; another, the Kokomo ditch, would extend 36.1 miles. These proposals never got beyond the filing stage. (1)

The idea of bringing water from the western slope to the eastern slope did not die, though. As early as 1913, the City and County of Denver began buying water rights west of the Continental Divide. A May, 1914, report spoke of the earlier proposals. A May, 1920, summary reviewed the points of the 1914 report. W.F.R. Mills, a former Denver mayor and, at that time, a manager for the Board of Water Commissioners, wrote an April 12, 1921, letter calling for the designing of a water diversion project on the western slope. (2)

On May 13, 1923, the State Engineer filed plans for a Blue River water diversion project, a proposal based on the Goldsborough and Bancroft plan as modified April 10, 1910. (This plan also formed the basis of the 1914 report.) The proposal was considered unfeasible and too expensive. (3)

The Board of Water Commissioners (also known as the Water Board) filed an alternate proposal with the State Engineer on October 19, 1927. (Created in 1918, the Denver Board of Water Commissioners is a five-person non-political board that has complete control of the water system supplying the City and County of Denver.) The October 19 plan featured a 23.3-mile-long tunnel with the western portal at an elevation of 8,845 feet and a low dam across the Blue River between the mouths of the Ten Mile Creek and Snake River.

Hoping to save money, the Board of Water Commissioners in 1932 proposed working with the Bureau of Reclamation to construct the tunnel. The Works Progress Administration (WPA) gave $100,000 to the Bureau of Reclamation for conducting a preliminary investigation. When the Bureau of Reclamation suggested using the Moffat Tunnel instead of building a new tunnel, the Board of Water Commissioners refused to consider the idea. With another grant, this time for $75,000, from the federal government, the Bureau developed a second proposal that the Water Board subsequently rejected. On December 31, 1941, both the Board and Bureau received additional grants to continue the study. Although preliminary studies had begun in 1913, it was not until November 14, 1942, that location plans were finally filed with the State Engineer. (4)

Progress slowed during World War II. When concern focused on Denver water needs once again, an Engineering Board of Review, composed of the Chief Engineer for the Colorado Water Conservation Board, the Senior Engineer from the United States Bureau of Reclamation, a Denver consulting engineer, and the Chief Engineer of the Denver Water Department, recommended on February 16, 1946, that the Montezuma tunnel project be approved for consideration. (5)

The Denver Water Board listed three purposes for building the dam: regulate stream flow, provide water storage, and act as a diversion facility taking water from the streams to the tunnel. (6) Initially, diversion was the only function of the dam but that changed when the Board of Water Commissioners determined that the dual functions of diversion and storage would prove more economical. By diverting and storing water, it would be possible to control the water flow for a hydroelectric plant should one be considered later. (In 1987, workers installed below the dam a 1.7 megawatt generator that feeds power to the Public Service Company grid.) (7)

Geology of the Reservoir Site

IN DETERMINING THE LOCATION OF THE DAM, the Water Board considered geologic structures at and below the surface, but geography overruled geology. The confluence of the Blue River valley north of old Dillon, the large basin for water storage, and the ability to maintain gravity flow through the Roberts Tunnel dictated the dam site despite less-than-ideal geologic conditions.

A geologic survey completed by Ogden Tweto of the United States Geological Survey (USGS) in 1945 revealed extensively fractured rocks that would need treatment to reduce permeability. He recommended that the dam be built from First Hill to Fairview (Lake) Hill rather than Placer Hill, just north of Fairview (Lake) Hill, because the Dakota sandstone/quartzite and Pierre shale comprising Placer Hill were not strong enough to serve as an anchor for the dam, while the sandstone/quartzite and Pierre shale of Fairview (Lake) Hill were. (8) *(Figure 6)*

A number of sedimentary rock formations underlay the reservoir site: from top to bottom, the Pierre, Niobrara, Dakota, Morrison, and Maroon. Precambrian schist, a metamorphic rock over 600 million years old, underlies sedimentary layers

Figure 6. Bridge over the Blue River, North of the Dam Site. First Hill, the eastern anchor for the dam, rises just behind the buildings in the background. (Courtesy Board of Water Commissioners, City and County of Denver)

of conglomerate, sandstone, mudstone, siltstone, and shale. (See *Roadside Summit, Part I, the Natural Landscape*, a Summit Historical Society publication for more about these rocks and rock formations.) The thinly bedded and very weak shale of the Pierre Formation outcrops in many areas throughout the county. The Niobrara limestone, really a calcareous (calcium-rich) shale, breaks easily. The resistant, hard, but brittle, Dakota Formation composed of sandstone and quartzite (metamorphosed sandstone) averages 225 feet thick. Much in evidence along the roads around the reservoir, the rock moves and faults easily. Below the Dakota Formation is the weak but impervious Morrison Formation known for its wealth of dinosaur bones. The impervious nature of the rocks of this formation offered a definite advantage in controlling water seepage. The Maroon Formation, whose red rocks can be

Figure 7. The Cutoff Trench. Dug parallel to the long axis on the water side of the dam, the trench was 30 feet wide at the base and 90 feet deep. The view looks east toward First Hill. (Courtesy Board of Water Commissioners, City and County of Denver)

Figure 8. Cutoff Trench. Workers have filled the narrow trench extending the length of the large cutoff trench. (Courtesy Frisco Historic Park & Museum and Polhemus Family Collection)

seen between Copper Mountain and Vail Pass, along Boreas Pass Road, and south of Breckenridge offered a good foundation for the dam. The glacial gravel extending between four and five miles along the western side of the reservoir site supplied a good portion of the fill needed to construct the earth-fill dam. (9)

At the dam site, the high permeability of the gravel extending between 20 and 80 feet above bedrock required that a cutoff trench be dug down to bedrock along the long axis of the dam parallel to the water-facing side of the structure. With a one-to-one side slope ratio, the trench measured

Figure 9. Impervious Material. Fifteen-ton Euclid dump trucks carried impervious material from borrow pits to fill the cutoff trench. (Courtesy Board of Water Commissioners, City and County of Denver)

30 feet wide at the base and had a maximum depth of 90 feet. A narrow trench extended down the middle of the large trench. At intervals along the bottom of the narrow trench, workers bored holes 300 feet deep and filled them with concrete. Then the workers filled the narrow trench with concrete. To prevent leaking water from undermining the earthen dam, the large trench was backfilled to ground level with impervious clay brought from near the west portal of the Roberts Tunnel. (10) *(Figures 7, 8, 9)*

Core samples taken at both the east and west abutments (First Hill and Fairview/Lake Hill) indicated that the rock layers here, too, needed strengthening. Workers dug long trenches measuring four feet deep by six feet wide into the easily fractured sedimentary rocks. They then drilled holes 200 feet deep and filled them with grout before filling the entire trench with grout. *(Figures 10, 11)*

Figure 10. Grout Cap on West Abutment. To provide added strength to the east and west anchors of the dam, grout caps, six feet wide and four feet deep, cut across the sedimentary rock layers. (Courtesy Board of Water Commissioners, City and County of Denver)

Figure 11. Filled Grout Cap on West Abutment. When filled with grout, the caps added stability to the easily fractured sedimentary layers. (Courtesy Board of Water Commissioners, City and County of Denver)

The High Dam

THE DENVER WATER BOARD considered three construction plans: a low diversion dam with a crest elevation of 8,896 feet and crest width of 32 feet; a high dam with a crest elevation of 9,031 feet and width of 32 feet; and the same high dam with a crest elevation of 9,031 feet and width of 72 feet to accommodate the proposed Interstate-70. (11)

The Water Board hesitated for many months before finally choosing the high dam with crest width of 32 feet. Financial considerations partially caused the delay. Only when support came from the Denver Chamber of Commerce and other civic groups did the Water Board make its final decision. The width of 32 feet precluded building Interstate-70 on the dam itself but provided the extra two-foot-width required for safety considerations. A vote by the commissioners on May 10, 1960, was unanimous and affirmative. Bonds would finance the additional construction costs. (12)

Bids for construction had been advertised, opened, and awarded before the Board made the final decision to build the high dam. The winning bid for the low dam came from Potashnick Construction, Inc., of Cape Girardeau, Missouri,

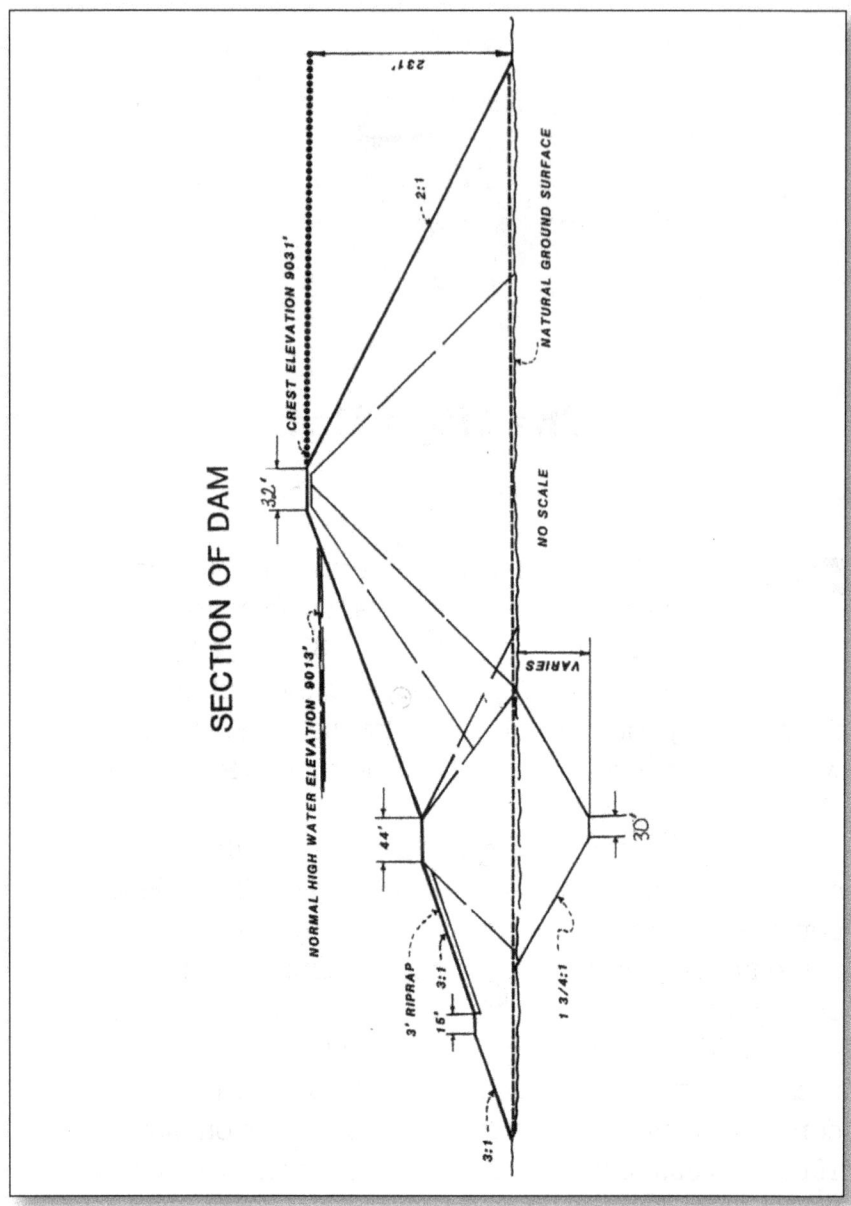

Figure 12. Cross-section of Dillon Dam. Made of 12 million tons of earth-fill, the dam rises 231 feet above the ground surface and extends 5,888 feet in length. With a crest width of 32 feet, the dam measures 2,700 feet at the base. (Redrawn from Diagram, Courtesy Board of Water Commissioners, City and County of Denver)

Figure 13. *Dillon Dam with the 15-foot Fixed Wing Gate and Glory Hole in Center Background. Although not of the best quality, the photograph confirms the shape of the dam as drawn in Figure 12. (Photographer Unknown)*

on October 27, 1959. The modified contract of June 4, 1960, included a dam 135 feet higher than originally planned and specified an expected completion date of December 31, 1963. (13)

The dam itself would consist of three sections: a trapezoidal-shaped central section extending from valley floor to crest formed from the same impervious clay-type material used to fill the cutoff trench and triangular sections of a pervious material flanking the core on both sides that would offer greater stability during downdraw. (14) ***(Figures 12, 13)***

The 231-foot-tall earth-fill dam would impound 252,678 acre-feet of water. An acre-foot of water (about the size of a football field) equals 326,000 gallons of water—enough to supply a family of four for one year. The dam on the upstream face would have a 2½-to-1 slope. This means for every one foot of horizontal distance, the face would rise 2½ feet. (15) ***(Figure 14)***

Figure 14. Northern Face of Dillon Dam. The face of the dam, with its 2½-to-1 slope, rises 2½ feet for every one foot of horizontal distance. The Glory Hole appears in the background. (Courtesy Board of Water Commissioners, City and County of Denver)

When the 12 million tons of earth-fill were in place, workers would cover the upstream face of the dam with a three-foot-thick layer of large boulders called riprap from the crest to an elevation of 8,440 feet to protect the structure from water and ice erosion. (16)

Clearing for the dam and borrow areas began on April 22, 1960. (17) A number of borrow areas had been designated, the largest areas just south of the new local Denver Water Board Headquarters along U.S. Route 6 and north of Frisco on part of the Giberson ranch. These borrow areas provided much of the 12 million tons of material for the earth-fill dam. They also supplied an on-site batch plant with aggregate for concrete. Trains brought the cement to Leadville; trucks carried it the rest of the way to the construction site. (18)

Workers cleared timbered areas to an elevation of 9,017 feet plus an extra 20 feet horizontally or four feet vertically. They removed brush to an elevation of 9,018 feet. Clearing the willows presented quite a challenge. Because their supple trunks and branches bent when pushed by the blade of a bulldozer, the willows had to be pulled out by the roots. Dirt clinging to the roots prevented burning so workers resorted to burying the vegetation. When this became a time-consuming job, the plants were killed with a mixture of 2-4-D and diesel oil sprayed from an airplane on July 21, 1962. The dead brush could then be burned. (19) *(Figures 15, 16)*

Three brush fires "escaped." Two on the north side of the Snake River burned about six acres of Denver Water Board land. Another, this one on the south side of the Snake, burned about 14 acres of Denver Water Board and U.S. Forest Service land. (20)

Figure 15. Tree Removal. Workers cut trees to an elevation of 9017 feet plus four feet vertically or 20 feet horizontally. (Courtesy Board of Water Commissioners, City and County of Denver)

Figure 16. Stacked for Burning. Workers removed vegetation up to an elevation of 9018 feet. (Courtesy Board of Water Commissioners, City and County of Denver)

Before the Blue River could be diverted around the construction site, an outlet works with "Glory Hole" had to be built. The outlet tunnel, 1,700 feet long and 15 feet in diameter, was partially cut into stable rock on the west side of the dam. The Glory Hole, the overflow outlet at an elevation of 9,017 feet, leads to a 233-foot-deep shaft that bends 90 degrees from vertical to horizontal. Maximum discharge is 15,000 cubic feet per second if the lake level reaches 9,025 feet—eight feet above the top of the Glory Hole. (21) Separate from this, but leading into it, is the inlet structure, which includes a 15-foot, six-inch fixed-wheel gate mounted horizontally. The gate takes water from the deepest and coldest part of the reservoir and feeds it into the 1,700-foot-long tunnel. It is water from the inlet gate that keeps the Blue River flowing. Anaerobic conditions might develop if water is not constantly removed from the coldest and deepest parts of a reservoir. *(Figures 17–25)*

Figure 17. Glory Hole. The Glory Hole, at an elevation of 9017 feet, is the over-flow outlet for the reservoir. (Courtesy Board of Water Commissioners, City and County of Denver)

Figure 18. *Glory Hole Excavation. Workers cut the 15-foot-diameter outlet works into stable rock on the west side of the dam. (Courtesy Board of Water Commissioners, City and County of Denver)*

Figure 19. Glory Hole Framework. The wooden framework provided added stability to the rock walls. (Courtesy Board of Water Commissioners, City and County of Denver)

Figure 20. 90 Degree Bend. The water drops 233 feet before making a 90 degree bend to the north. (Courtesy Board of Water Commissioners, City and County of Denver)

Figure 21. Outlet Works. Not all of the outlet works could be cut through stable rock. When complete, this part of the outlet works had to be covered with earth-fill. *(Courtesy Board of Water Commissioners, City and County of Denver)*

Figure 22. Outlet Tunnel Interior. Workers installed wooden and steel supports before lining the tunnel with highly polished concrete. *(Courtesy Board of Water Commissioners, City and County of Denver)*

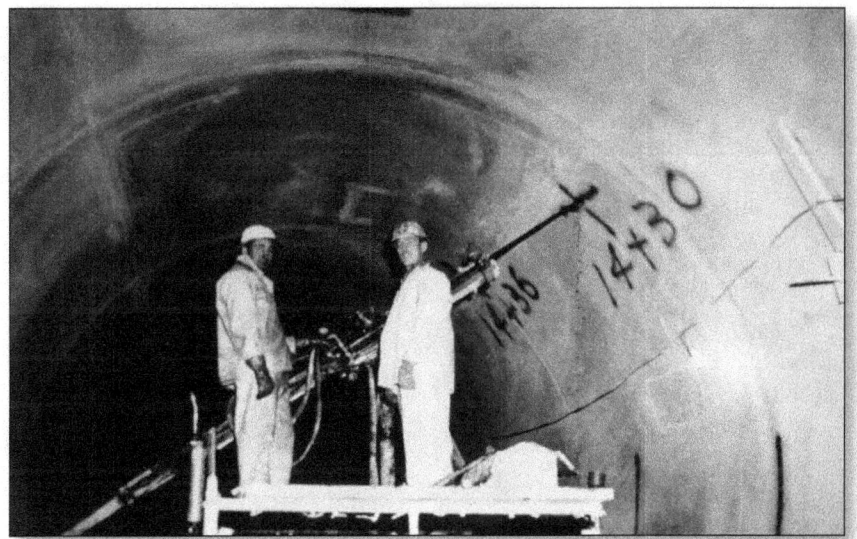

Figure 23. Low Pressure Grout. Men drilled holes in the tunnel lining in order to force low pressure grout into weak areas. (Courtesy Board of Water Commissioners, City and County of Denver)

Figure 24. Reinforcing Steel. Layers of reinforcing steel provide strength to the base slab at the outlet. To accommodate a maximum discharge of 15,000 cfs, several layers of "rebar" were required. (Courtesy Board of Water Commissioners, City and County of Denver)

Figure 25. Horizontal Fixed Wing Gate. The fixed wing gate (center of the photograph) maintains the flow of the Blue River. Water moving through the gate comes from the deepest, coldest portion of the reservoir thus preventing anaerobic conditions from forming. (Courtesy Board of Water Commissioners, City and County of Denver)

Contractors developed a unique design for the Glory Hole, building and testing a plastic model of it at Colorado State University's hydraulic laboratory before installation began on site. (22) The four large "fins" assure that water drops straight down into the Glory Hole, rather than swirling. Air bubbles trapped in swirling water would "pop," causing damage to the concrete lining. *(Figures 26, 27, 28)*

In order to prevent houses from being built over the water or sewage systems near the water, the Water Board announced plans to purchase or condemn a buffer strip around the lake. The strip would include the first 25 feet above water level. Thus with a planned water level of 9,025 feet, the Water Board would control all land between 9,025 feet and 9,050 feet along the shore. (23) This also meant that there would be no public property on the lake shore.

Figure 26. The "Fins" of the Glory Hole. Scientists in the hydraulic laboratory at Colorado State University designed the Glory Hole. (Courtesy Board of Water Commissioners, City and County of Denver)

Figure 27. Role of the Fins. The fins prevent water from swirling as it enters the Glory Hole. Swirling water would allow popping air bubbles in the water to wear away the concrete of the Glory Hole in a process called cavitation. (Courtesy Frisco Historic Park & Museum and Polhemus Family Collection)

Some county residents did not welcome these plans. In an article appearing August 5, 1960, the writer called it another Denver Water Board "land grab" that would take even more land with development potential off the tax rolls. The reporter reminded the readers that the strip would not be 25 feet wide but would vary in width to include 25 feet in elevation. If a stretch of land gained elevation slowly, the strip would be very wide indeed. The reporter noted that much of Frisco lies between 9,025 feet and 9,050 feet in elevation. (24)

The Water Board responded by saying that the buffer strip was intended to keep the water clean—a prime concern for those entrusted with supplying water to the City and County of Denver. In a statement directly counter to an earlier one proposing a recreational role for the lake, the Water Board claimed that public access to the water was not a necessity

Figure 28. *Protecting the Glory Hole in Winter. To prevent ice from forming and eroding the concrete Glory Hole, workers drop this covering in place in the fall. (Courtesy Frisco Historic Park & Museum and Polhemus Family Collection)*

because the reservoir was not designed as a recreation site. Many remembered that an argument for the high dam had been the resulting recreational possibilities. (25)

A few weeks later, the Water Board agreed that the water behind the dam would be available for recreational purposes. (26)

To clarify the issue, the Water Board told residents it intended to obtain just some of the land between elevations 9,025 feet and 9,050 feet for the buffer. The Board set a limit of 300 horizontal feet from the water line. No horizontal limit had been set previously. (27)

The Water Board did not follow those guidelines precisely, though. The Board sold some land on the Dillon waterfront below 9,025 feet, while condemning some land at the same elevation on the Frisco side. (28)

Three unforeseen problems surfaced during construction: two much easier to overcome than the third. For greater stability, the dam itself had to be extended farther east than anticipated by early surveys. Obtaining the necessary fill required a greater amount of excavation in the borrow area. These two problems were minor compared to the third. Sliding of the easily fractured rock on the wall of the west abutment created a "recurring and urgent problem." (29) The unstable rock, located outside of the construction zone, continually moved downslope threatening the outlet works and building. Trucks hauling fill along a road cut in the west abutment aggravated the situation during the 1961 construction year. When heavy equipment used the road in 1963 for topping out the dam, movement of the material began anew. (30) *(Figures 29, 30, 31)*

Figure 29. Protection for the Glory Hole. When huge trucks loosened rocks on the steep wall of the west abutment, it became necessary to protect the Glory Hole and building standing next to it. (Courtesy Board of Water Commissioners, City and County of Denver)

Figure 30. Lodgepole Pine Trunks. *A covering of lodgepole pine trunks protected the building from the falling rocks. (Courtesy Board of Water Commissioners, City and County of Denver)*

Responding to concerns of those driving the road, the county commissioners declared the west abutment road unsafe for vehicular traffic. (31)

A wire mesh hung over the disintegrating rock wall failed to solve the problem. An article appearing in the *Summit County Journal* aptly tells the story.

"One of the most scenic roads in Summit, the dam road has recently blossomed into the county's biggest transportation headache. And county officials, wary of spending a small fortune on a non-essential road, aren't yet sure how to cure it. Flat, wide, and gently curved, the dam road runs into trouble just west of the dam. At that point, a steep slope some 200 feet high periodically unleashes a freight of rock fragments onto the road. Plummeting rock is especially likely when the slope is loosened by spring snowmelt or heavy rains.

Figure 31. Wire Mesh on the West Abutment. Workers covered the west abutment with wire mesh to protect vehicles using the road below. (Courtesy Board of Water Commissioners, City and County of Denver)

A wire mesh designed to catch tumbling rocks covers the slope's lower half, but the rocks are starting to uproot this protective device. That means more rocks make their way to the road with less effort.

The slope is cohesive enough not to suddenly crumble and shoot into the lake, but it's also loose enough to continue its deterioration and cause increasing rockfall.

The slope, constructed by the Colorado Division of Highways when the dam was built in the early '60s, is far too steep to support the sandstone and shale which make it up. And the longer nothing is done about it, the worse the situation will get. Shale beds erode easily crumbling more the longer they've weathered." (32)

The Roberts Tunnel

TO CARRY WATER UNDER THE CONTINENTAL DIVIDE, the Water Board planned to build the longest major underground tunnel of its kind in the world. Construction began on September 17, 1942. After a hiatus during World War II, work resumed on June 24, 1946. (33) *(Figure 32)* The tunnel had several names. First called the Blue River Tunnel, then the Montezuma Tunnel, it officially became the Harold D. Roberts Tunnel, named for a Denver Water Board attorney instrumental in buying water rights on the Blue River and gaining Congressional approval for the project. (34)

To assure gravity flow, the tunnel's west portal has an elevation of 8,844 feet, 174.24 feet higher than the eastern portal. The gradient averages seven feet per mile. This elevation placed the tunnel entrance below the Blue River's channel. (35) With an inside diameter of ten feet, three inches, the tunnel can carry 1,000 cubic feet per second (680 million gallons per day) if the water elevation is 9,017 feet. Water exiting the tunnel enters the North Fork of the South Platte River along U.S. Highway 285, about one mile west of Grant.

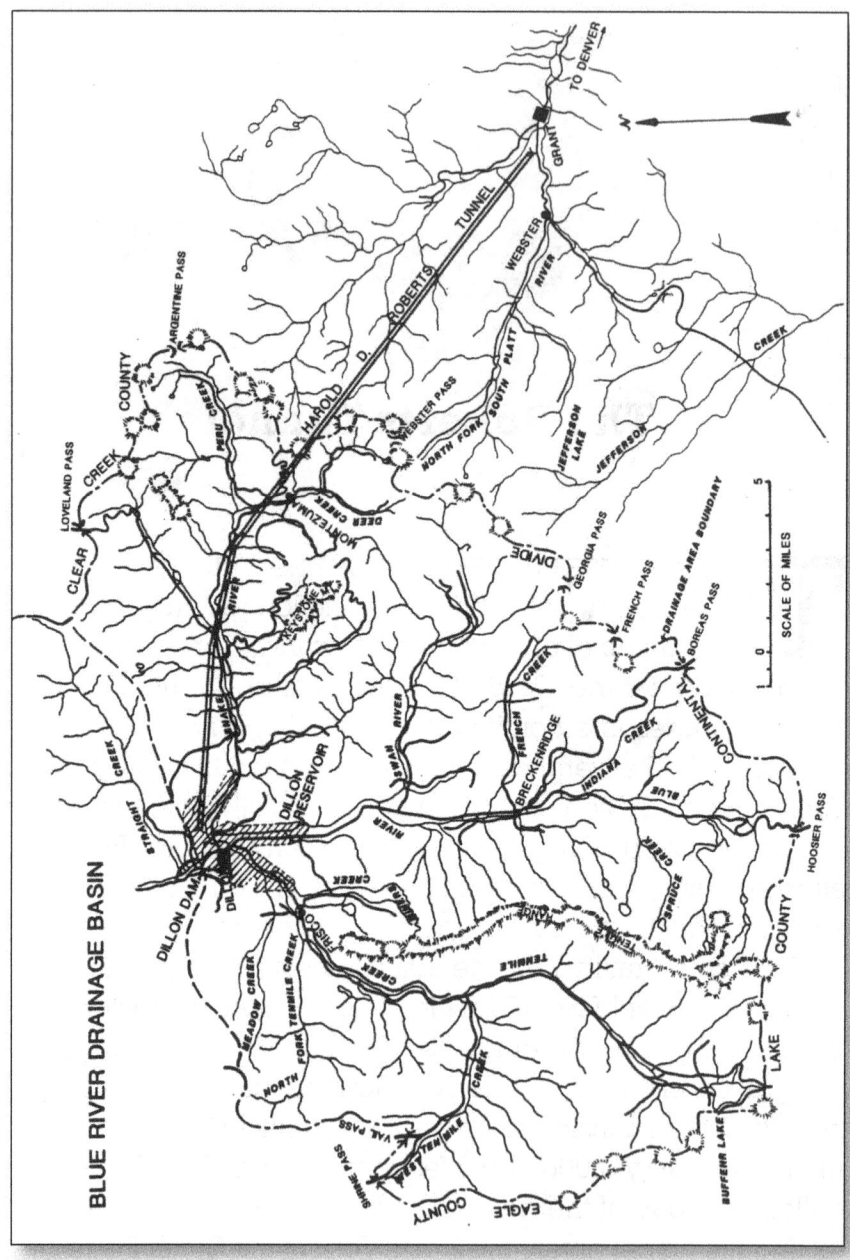

Figure 32. *Blue River Drainage Basin. The Roberts Tunnel stretches 23.3 miles from Dillon Reservoir to Grant. (Redrawn from Map, Courtesy Board of Water Commissioners, City and County of Denver)*

Geology along the Roberts Tunnel

THE CONSTRUCTION OF THE ROBERTS TUNNEL provided geologists with the opportunity to learn more about the geology of Summit County. Their studies found rock formations along the tunnel route similar to those found throughout the county: Precambrian metamorphic and igneous rocks along the eastern two-thirds of the tunnel and near the Williams Fork Thrust Fault; Paleozoic and Mesozoic sedimentary rocks at the western end of the tunnel; and Cenozoic igneous rocks near the town of Montezuma where the tunnel cuts the Montezuma stock. (See *Roadside Summit, Part I, the Natural Landscape* for complete explanations of these terms.) The rocks tell of compressional uplift during the Precambrian and Cenozoic eras—the times when the Ancestral and present-day Rocky Mountains formed. Granitic rocks forced their way into the 600-million-year-old gneiss (pronounced "nice") and schist here as in the Ten Mile and Gore ranges when the forces creating the Ancestral Rockies metamorphosed the pre-existing rocks into gneiss and schist. The sedimentary rocks cut by the tunnel include the Maroon and Minturn formations—the same formations found at Vail Pass and forming Jacque Peak, Red Mountain, Mount

Argentine and others south and east of Breckenridge. The highly fragmented Dakota sandstone cut by the tunnel is the same formation serving as an anchor for the east abutment of the dam. (36) *(Figures 33, 34, 35)*

The tunnel intersects the Williams Fork Thrust Fault at two places. The thrust fault developed during the episode of crustal compression that created the present Rocky Mountains. Land east of this 60-mile-long fault moved west about one mile, superimposing 600-million-year-old metamorphic and igneous rocks on top of much younger sedimentary rocks. Gravity and the layers of slippery shale under the thrusting Precambrian rocks enhanced sliding. (37) These sedimentary rocks of the Pierre (pronounced "peer" by geologists) Formation line U.S. Route 6 between Dillon and Keystone and Colorado Route 9 to the east of Green Mountain Reservoir.

While the Rockies were forming, magma (molten rock) pushed into horizontal and vertical cracks in pre-existing rocks and solidified, creating sills and dikes. Highly mineralized masses of magma also melted their way upward but didn't quite reach the surface. When cooled, these masses, called stocks, take on a shape somewhat like hot air balloons. The highly mineralized Montezuma stock, measuring 16 square miles at the surface, is the largest in the Front Range of the Rocky Mountains. (38) Some of this magma experienced a two-stage cooling process, creating a porphyritic texture. (39) Slow cooling far below the surface allowed large crystals to grow. Subsequent uplift with more rapid cooling followed and a fine crystal matrix formed between the larger crystals.

The first surveys and core drillings along the tunnel route began July, 1931, and engineers responded to their findings. When they learned that the rock near the Williams Fork Thrust Fault was so highly fractured, they moved the west portal north 3,500 feet. (40) Wanting to take advantage of

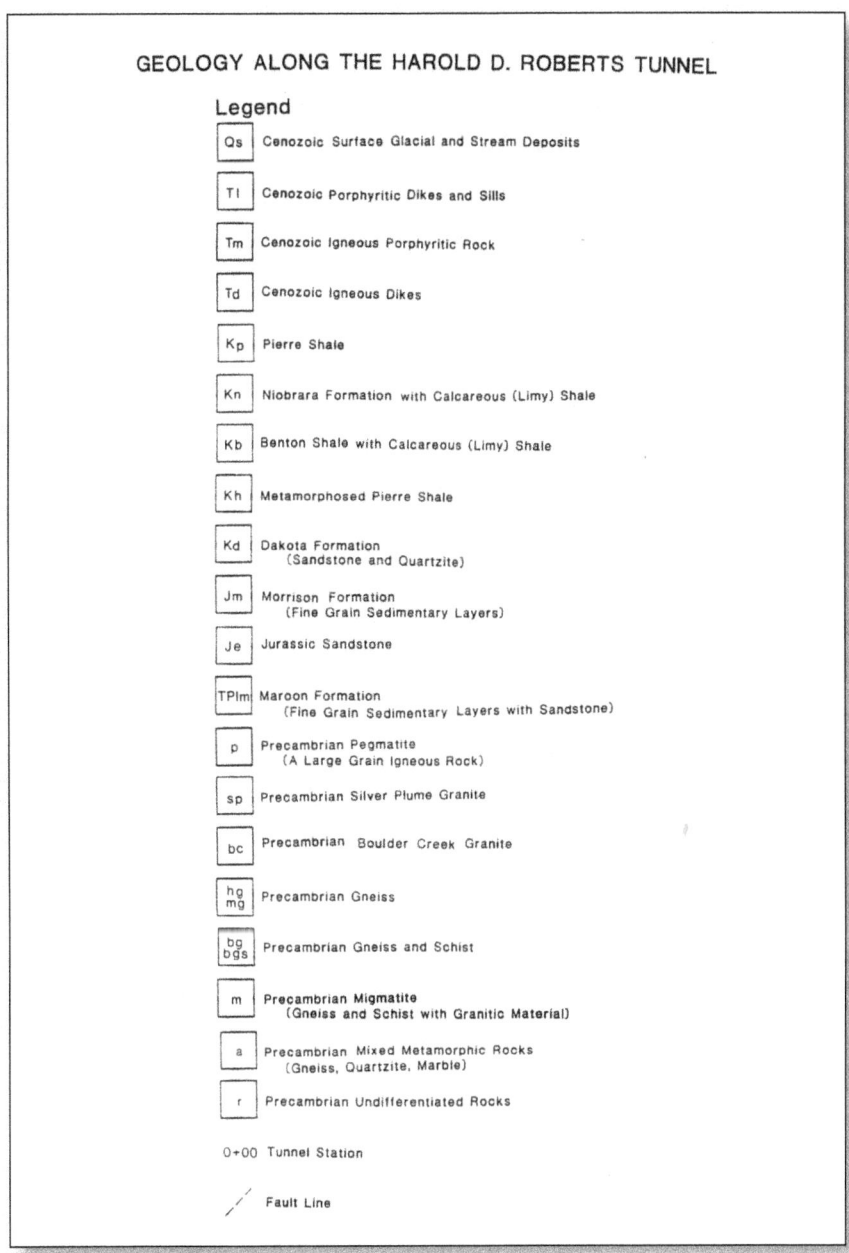

Figure 33. Geology along the Harold D. Roberts Tunnel: Legend. (Adapted from geologic map and section along the Roberts Tunnel Line, Park and Summit Counties, Colorado. United States Geological Survey Professional Paper 831-B)

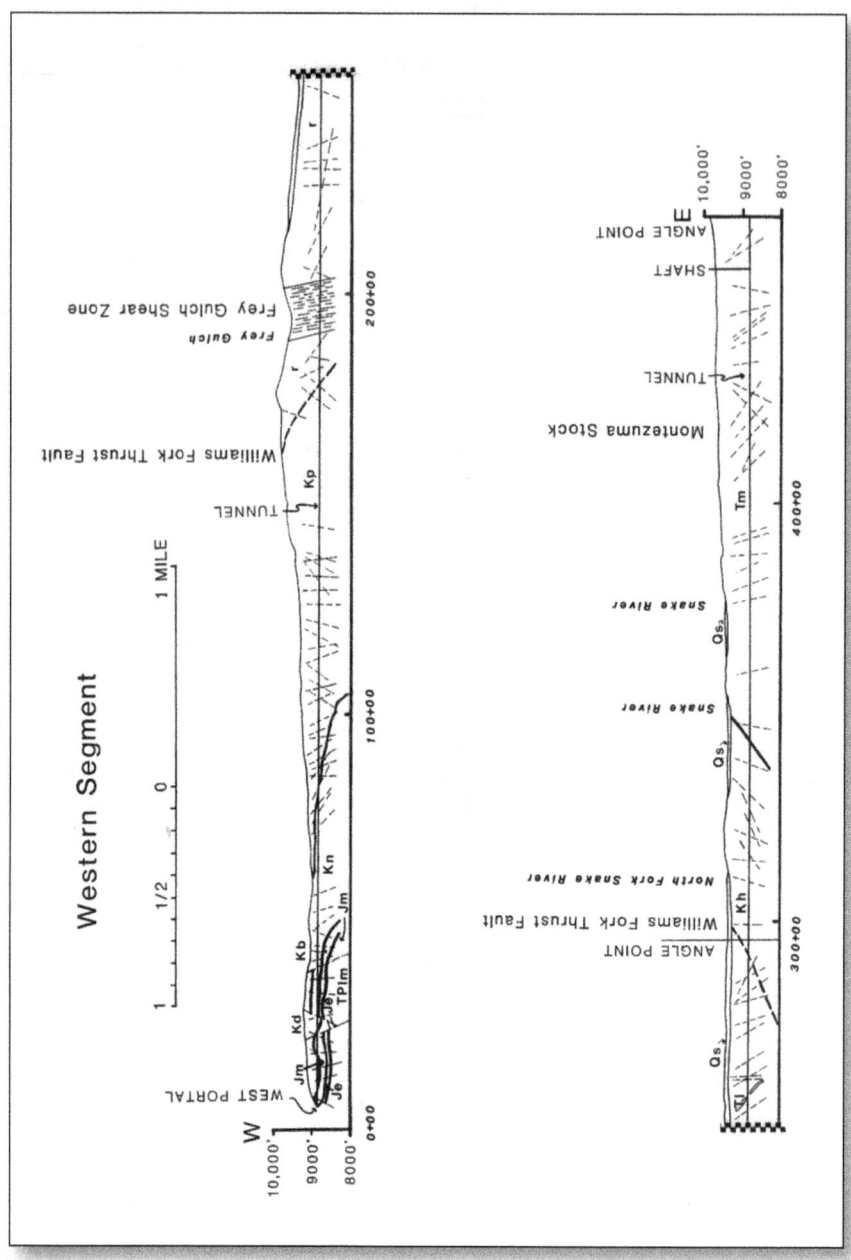

Figure 34. Western Segment. (Adapted from geologic map and section along the Roberts Tunnel Line, Park and Summit Counties, Colorado. United States Geological Survey Professional Paper 831-B)

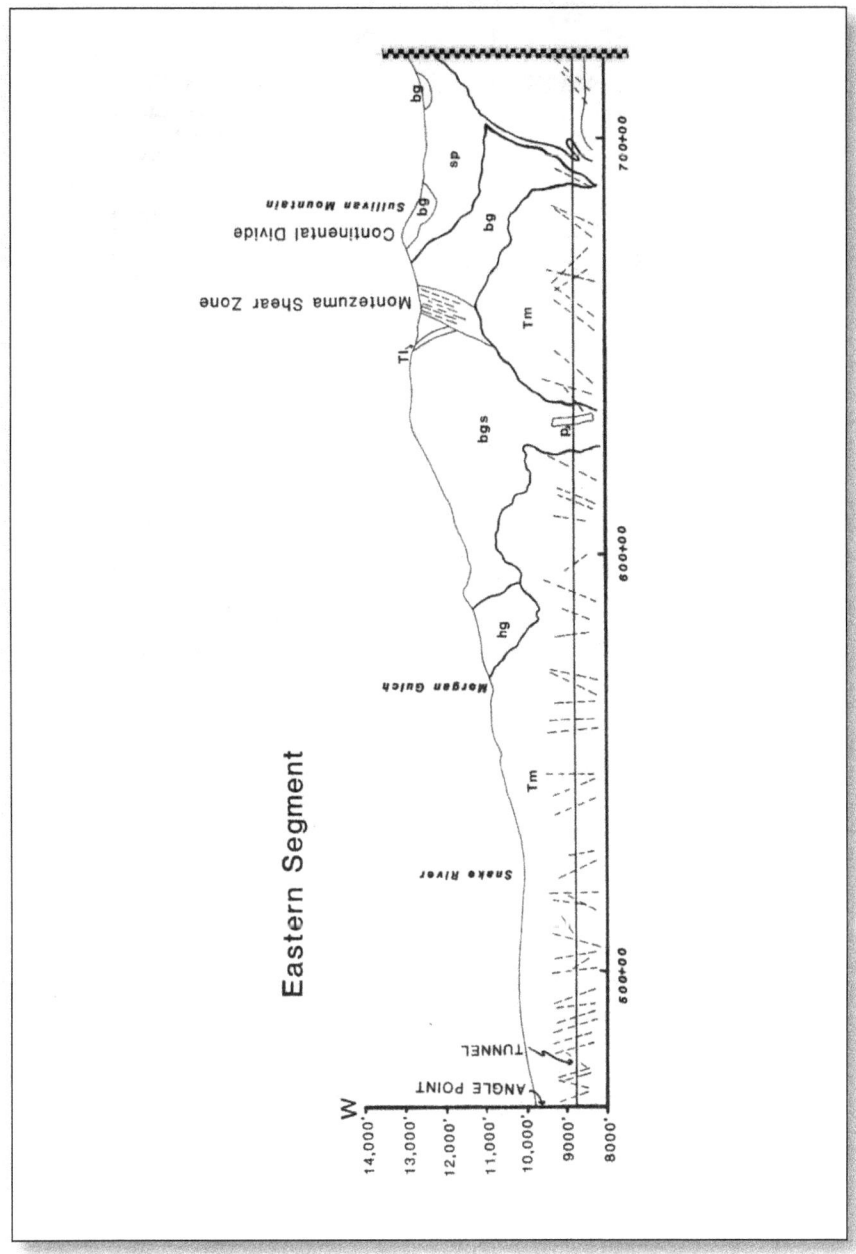

Figure 35a. Eastern Segment: Section 1. (Adapted from geologic map and section along the Roberts Tunnel Line, Park and Summit Counties, Colorado. United States Geological Survey Professional Paper 831-B)

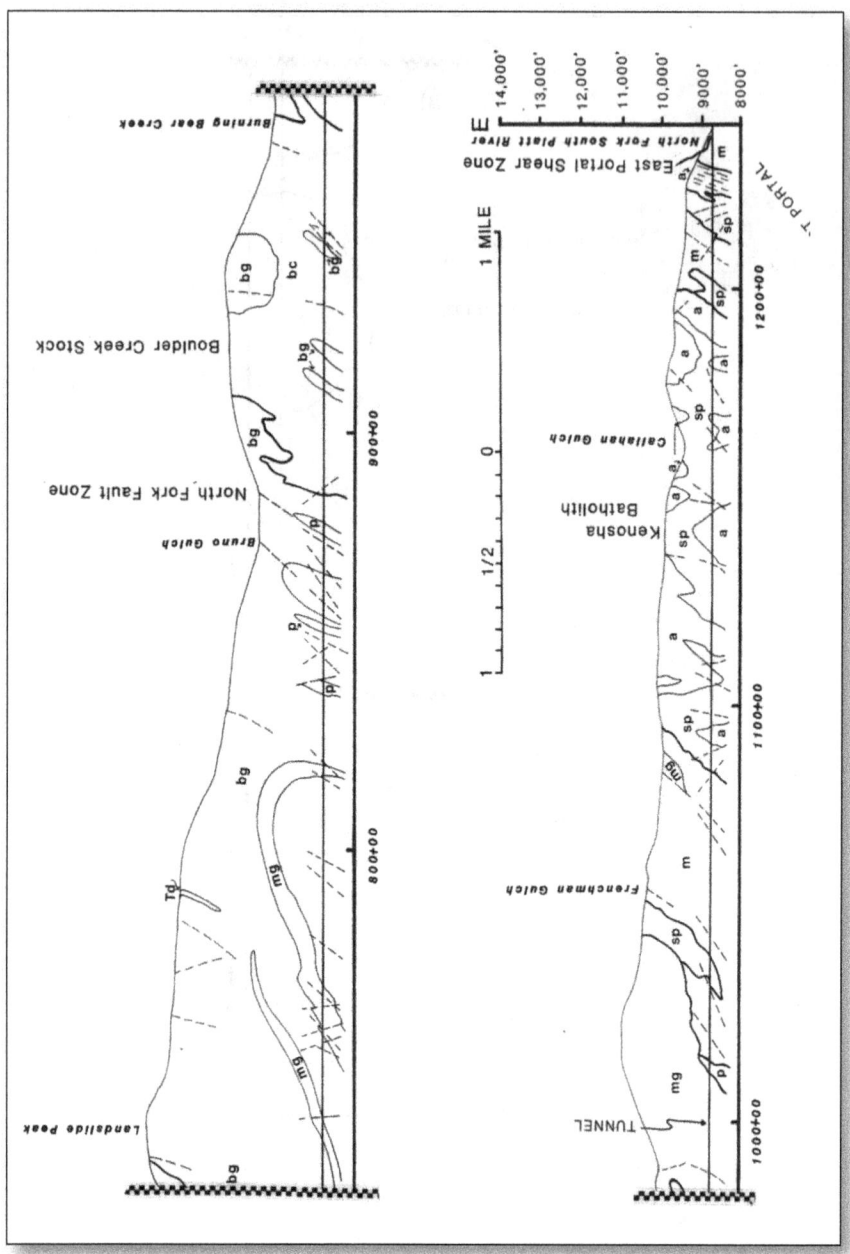

Figure 35b. Eastern Segment: Sections 2, 3. (Adapted from geologic map and section along the Roberts Tunnel Line, Park and Summit Counties, Colorado. United States Geological Survey Professional Paper 831-B)

the strength of the granite in the Montezuma stock, the engineers designed a bend in the tunnel's path. This eliminated the need to bore through the highly fractured rock that a straight-line route would have encountered. (41) Of the 23.3 miles of boring, workers cut through 3.23 miles of folded sedimentary rocks, principally near and below the Williams Fork Thrust Fault; 2.11 miles of shattered, fractured metamorphic rocks such as gneiss and quartzite; 1.06 miles of brittle, baked shale that required much high pressure grouting; 6.4 miles of jointed, fractured igneous rocks that required steel supports; and 10.5 miles of Precambrian rocks including schist, gneiss, and granite. Only 57 percent of these Precambrian rocks required steel supports and timbering whereas 91 percent of the rest of the tunnel did. Altogether 72 percent of the tunnel required steel supports and timbering. (42)

With the final route cutting across the proven Montezuma mineral belt, it is surprising that few veins of commercial value were found during tunneling. The minerals discovered came mainly from the highly mineralized Montezuma stock and nearby Precambrian granite, gneiss, and schist. (43)

Building the Roberts Tunnel

D IGGING THE DIVERSION TUNNEL PROVED DIFFICULT. Workers drilled dynamite holes, loaded the charges, shot the dynamite, mucked out (removed) the shattered rock, inserted steel supports, timbered, drilled feeler holes to determine the amount of grout (cement) required, and pumped in the grout. This process they repeated hundreds of times before they finished the tunnel. (44)

In some areas, the strength of the rocks required tungsten carbide drill bits. Dynamite came from DuPont, a company that supplied dynamite to Summit County mines in the late 1800s and early 1900s. A machine drilled 46 holes at a time. For every 20 feet they advanced in the tunnel, the men used 20 pounds or more of dynamite. (45) Each 200 to 250 pounds of dynamite detonated created 34 cubic yards of muck. Huge fans removed the dust and gases created by the explosions. When reversed, the fans blew 10,000 cubic feet of fresh air per minute into the tunnel. (46)

An air-driven, track-mounted shovel did the mucking. Diesel or electrically driven locomotives running on 56-pound (per yard) rails spaced 24 inches apart pulled the rock-filled cars from the tunnel. (47) Bought as salvage, the rails had been part of a Burlington-Northern spur in Nebraska. (48)

In some areas, the fractured, unstable rock walls required timbering before the mucking process could be completed. (49)

Just as miners encountered large amounts of water in some of their tunnels, water in large amounts from the Snake River flooded the diversion tunnel. Workers forced cement into the water-logged rocks to stem the flow. (50) Steel ribbing prevented the collapse of water-soaked, unstable walls. Even so, two pumps, each with a 2,500-gallon-per-minute capacity, worked continuously to remove the water. Some days, the pumps returned nearly 3.5 million gallons of water to the Blue and Snake rivers. (51)

In order to speed construction, workers dug a vertical shaft 910 feet deep near Montezuma. From there, workers blasted east and west to meet the tunnels being built from both the east and west portals. They reached a major milestone when they blasted through the last 12 feet of rock in the western end of the tunnel on January 2, 1960. The alignment of the two parts was almost perfect—only about one inch off. (52)

The eastern section opened February 24, 1960, when workers blasted through the last 45 feet of rock. (53) These two sections were also in almost perfect alignment.

With an outside diameter of 16 feet, the cost of the tunnel averaged $25 per inch or $300 per foot, for a total of $46 million in 1961 dollars. (54)

The pieces of tunnel lining, measuring 30 feet in length and formed of one-quarter inch highly polished steel plate, were inserted in two pieces. The lower section, forming an arc of 80 degrees, became the floor and the larger, upper section, which formed an arc of 280 degrees, became the ceiling and two side walls. (55) A low pressure grout (100 pounds per square inch) of sand and cement filled the space between the rock walls and liner. (56)

Although the tunnel was completed May 7, 1962, water first flowed through it on July 17, 1964. (57)

Old Dillon

DURING THE DEPRESSION YEARS, the Denver Water Board gained control of much of the reservoir site by purchasing property at tax sales. By the 1950s, the Water Board owned as much as three-fourths of the land in town, making it impossible to ignore the fact that a dam would be built across the Blue River. *(Figures 36, 37, 38, 39)*

Representatives of the Denver Water Board and people in the Dillon area met at the Wildwood Lodge on November 3, 1955, to hear the Water Board's plans and to air their concerns. According to the plans at that time, the reservoir would impound 52,000 acre-feet of water and become a permanent recreation site. To lessen the trauma of losing homes and property that might have been owned by several generations of the same family, the Water Board noted that the reservoir, both during and after construction, would bring jobs and business opportunities to the county. A Water Board representative predicted that the proposed new town as well as the homes and businesses in it would be better than what existed presently. The Water Board offered residents the option of buying land in the new town or exchanging their property for some in new Dillon. It also offered to advance funds for the

Figure 36. Old Dillon looking South toward Swan Mountain. Throughout its history, Dillon served as a transportation hub, an agricultural supply town, and a watering hole for ranch hands. (Courtesy Summit Historical Society)

water and sewer systems in the new town that could be repaid at the time of the town's incorporation. In addition, the Water Board would allow property owners to move buildings and structures that the Water Board had paid for in negotiated settlements.

Local residents expressed strong feelings about a number of concerns. Senior citizens on fixed incomes would be displaced. They would not benefit from the business upswing and would not receive enough from the sale of their property to remain in the area. The influx of construction workers would bring an increased number of children to local schools. Who would supply the funds to hire new teachers and buy more classroom supplies? As the Water Board bought additional property, that acreage was removed from county tax rolls. How would the county continue to provide basic services with reduced financial resources?

Figure 37. Main Street, Dillon, in the 1950s. Dillon Drug in the foreground and other businesses lined the street. (Courtesy Rocky Mountain View Company)

Several residents expressed concern about who would decide the value of an individual property. Someone from Denver might not be sensitive to land values in Summit County—especially with someone whose family had occupied the land for several generations—whose ancestors might have been some of the original settlers of the county. Knowing of the plans to build a reservoir, many living within the construction site had not spent any money maintaining their buildings. Would that now work against them? Some buildings were too old and unstable to move. Would the money received be sufficient to construct new ones? Who would mediate when a difference of opinion arose about the appraisal of a property or building?

In all matters, the Water Board wished to work with one committee of residents who represented the majority view. The Water Board and county residents expressed this

Figure 38. *Dillon Humor. Yes, some people expressed their unhappiness. (Courtesy Frisco Historic Park & Museum and Polhemus Family Collection)*

theme—speaking with one voice—repeatedly. Some felt the lack of unity worked against land holders and expressed disappointment that the large land owners sold first, and sometimes eagerly, leaving the smaller owners to be "picked off" one at a time. (58) The Water Board noted that the lack of unity on the part of local residents slowed progress on deciding what the function and purpose of new Dillon would be.

Because so much concern remained about land owned by the Water Board being removed from local tax rolls, State Senator Fay De Berard introduced a state constitutional amendment requiring Denver to pay taxes on land owned outside city limits to the county where the land was located. Such tax-free land created hardships for county governments. Despite the support of legislators in the counties that would benefit from such an amendment, urban legislators doomed the proposal.

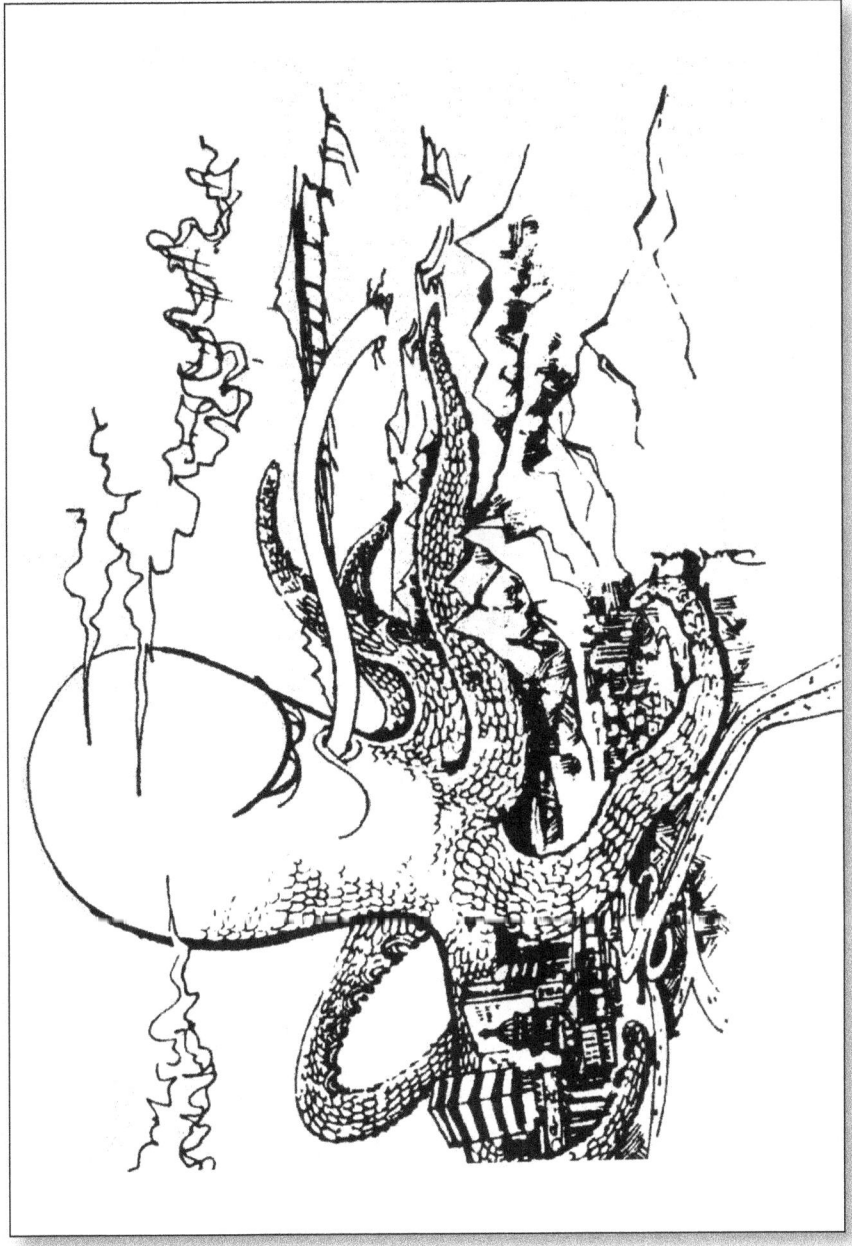

Figure 39. *Western Slurp. This cartoon appearing in the Summit County Journal expressed the feelings of many in the county. (Artist and Date Unknown)*

Responses to the Denver Water Board's desire to buy the remaining necessary property were mixed. Some resisted; some held out for a higher sale price. According to the editor of the *Summit County Journal,* some sold willingly, even approaching representatives of the Water Board long before the Board had a representative contact them. (59)

When work on the dam finally began, the Water Board owned or held options on almost all buildings and land needed in the Dillon area. (60)

For its permanent headquarters in Summit County, the Denver Water Board chose a part of the Byers ranch at the eastern end of Borrow Area #1. (61) When the Byers family sold its Tenderfoot Ranch to the Water Board, it was not the first time they had sold to a governmental agency. Part of their ranch in Pando, Eagle County, Colorado, became Camp Hale. This time, though, almost all of their ranch became property of the Denver Water Board. (62)

By the beginning of 1959, the Water Board was completing negotiations with the few residents still owning property within the site. Dillon hired an appraiser to determine the value of town property such as the town hall, water system, and sewer system. The Board hired an engineer to determine replacement costs for these facilities. The county commissioners did the same thing for county roads that would be inundated and need replacing. The town board suggested that school districts take the same steps. A seven-acre tract would be reserved in the new town for a school. (63)

By spring of that year, it seemed that maybe the trauma of leaving old Dillon and establishing new Dillon might still be a long way in the future. Until the Water Board decided the size of the dam and finalized the routes of the highways, plans for moving the town were premature. Possibly, construction would be delayed as much as 30 years if the amount of water needed by Denver could be obtained simply by using the diversion tunnel already under construction. (64)

For those opposed to the dam, the hope was short-lived. A letter mailed to Dillon residents in December, 1959, indicated that the old town must be evacuated by April 1, 1961. The school would remain open until June 1 of that year allowing the children to finish the school term. Most of the people could occupy their homes until that time. If construction required individual properties before then, the Water Board would notify those concerned. (65)

Even though people knew that eviction notices were as inevitable as the first snowfall each year, their arrival was traumatic and ill-timed. Those notified in August of a November eviction date faced the prospect of snow-covered ground on moving day. Although people hoped that roads in the new town site would be open to facilitate final evacuation, it quickly became evident that new Dillon would not be ready to receive its first inhabitants. The water and sewer systems, roads, lights, and electricity were not ready. Thus the newspaper reported on a variety of responses to these notices and others setting an August 1, 1961, final evacuation date. Some moved their buildings to the new town site but found living quarters elsewhere for a brief time. Some moved buildings to a temporary site and occupied them there until the building could be moved to new Dillon.

Hopes had been high for those wanting to move to new Dillon. But uncertainty about the site, highway relocation, and the future for businesses in the new town took their toll. Some decided not to accept the risks involved.

While some expressed optimism about the future of new Dillon, predicting the development of winter businesses because of the town's proximity to Arapahoe Basin and Loveland ski areas, others preferred moving their businesses to Silverthorne Flats along Colorado Route 9 where they hoped the atmosphere of old Dillon might be preserved. (66)

Businesses that could not open immediately in the new town faced extreme hardships; although for some, those businesses not profitable during the winter months, it seemed better to lose customers at that time rather than during the summer tourist season.

By January, 1960, 78 land-owners had already sold their property to the Water Board. Of those, all but 26 moved from the area. Of the 26, 12 expressed interest in resettling in new Dillon; only two of them definitely had plans to choose the new town. (67)

Throughout the next two years, the *Summit County Journal* reported on the movement of people and buildings out of the old town. Some went to Denver; some to Breckenridge, Keystone, Frisco, or the new town of Silverthorne. Some remained in the county; others moved beyond its borders. Some moved to apartments, trailer parks, new homes, or ranches. Some expressed bitterness; others resignation. Some could not afford the cost of a lot in the new town; others were sure the town would never be built.

Those choosing new Dillon included the Hank Emore family and the Mike Uitich family—both would move their homes from old to new Dillon. Kate Laskey planned to build a home in the new town. (68)

Charles Collard, an oil company bulk agent, told a *Summit County Journal* reporter, "I'm going to haul my house up there to the new town site and board up the windows. Next July I'll go there when water and sewer lines are connected and get it ready."

Mrs. Carl Prestrud, who with her husband operated a filling station, said, "We're going to give up our lease here November 1. Winter is not a profitable time here."

Faye Bryant, proprietor of the Arapahoe Café, expressed happiness with staying in business as long as he could, serving the 200-plus workers. Ultimately he would move his restaurant to the new town site.

Motel operator Thomas A. Foster had an option on a lot in the new town site as well as a house lot. "I don't see how the new town site can miss," he opined. "It will be right near U.S. 6 when they relocate it. We'll be closer to the Loveland Ski Area—and we're already doing fine with Arapahoe Basin. In summer the tourists are just like deer. They're all around."

John Sondregger, 60, a former grocer who leased out his building until the 30-day eviction policy made leasing difficult, expressed unhappiness with the new town site. "I can't afford to move to the new town site," he said. "Their lots cost too much. I'm gambling that there will be more business to be made on Colorado Highway 9 (the road to Kremmling) than there will be up there." His wife added bitterly: "Dillon is going to be scattered to the four winds."

Catering to people who sought something other than new Dillon was Virgil Cox, operator of the Blue River Inn. Cox promoted the Silverthorne development on Colorado Highway 9. Cox made no secret that he hoped to appeal to people on the grounds that the new town would be too fancy, too expensive, and, perhaps, too far from a highway for businesses. "We are trying to preserve the present town, or the look of the present town," Cox said. "We'll have well water and we'll have telephone and Public Service power within two or three weeks."

This appealed to people like the Sondreggers who liked Dillon the way it was. But John Bailey, head of the Dillon Improvement Corporation, added that "there's nothing wrong with the new Dillon plan. Lot sales are brisk. We've sold one-third of our lots, we plan to sell another third soon (from a waiting list) and we'll sell the rest next spring. But the price probably will be higher then."

The newspaper advised its readers about former Dillon residents. The Norman Ashlocks, some of the first to move from Dillon, went to Denver and opened a trucking business. The Lege family moved to a small acreage just outside of Denver, near Morrison. Karl Knudsen and his wife and grandparents, the Ed Rileys, moved to Wheat Ridge. Ada and Enair Lundgren and his mother, Hilda Cummings, bought a home in

Breckenridge. Minnie and Eric Ericson moved their home to Keystone. Roy Kohn moved his home and trailer court to Silverthorne. Lorenz Valaer moved the house he bought from Bessie (Warren) Blundell to Ivan Smith's ranch. Grace Warren and Bradish Warren also moved to the same ranch. Carl Prestrud and family now lived in Frisco. The Fritz Heagey family moved to Wyoming and bought a trailer court. Gay Loomis moved her home to Breckenridge and lived there. (69)

Condemnation proceedings began in January, 1960, against those owners holding property vital to the completion of the project. (70) By June 9, 1961, several properties were still needed: the Earl Rice ranch, the Howard Giberson ranch, the Merle Stewart and Walter Byron properties and Don Neet's Meadow Creek Ranch.

Business in old Dillon faced unique problems in 1960. Wildwood Lodge, owned by Harold Minowitz, and the Home Café, operated by Nora and Harry Calaghan, planned to curtail their business operations. Mrs. Prestrud said that Karl's Auto Service would close November 1 with hopes of being able to begin in some new location the next summer. Mike Vincent, of the Dillon Garage, planned to take his business to Tucson, Arizona, while Mr. and Mrs. Lester Adrian expected to move their theater and motel business to Frisco. The Ray Kohls hoped to operate a trailer court in Frisco and another in the Silverthorne subdivision. The Dillon Market would close for the winter but the Dillon Inn, Mint Pool Hall, and Elaine's Beauty Shop would open at new locations in the Silverthorne area. The Hamilton Hotel, which was moved to Breckenridge in the fall of 1961, burned May 15, 1963, while being remodeled. In addition to the fire companies in Breckenridge, the Fairplay Fire Company responded, reaching Breckenridge in 21 minutes after being told flames were consuming the entire main street in Breckenridge. (71)

The Dillon Drug Barbershop, Staghorn Liquor, Arapahoe Café, Lucky Horseshoe Bar, Center Grocery, Holiday House

Hotel, and Antler's Motel expected to continue in their present locations as long as permitted. The Lucky Horseshoe had no plans for the future. Fred Grantham of Dillon Drug, the John Valaers of Staghorn Liquor, Austin Otterson of the Antlers Hotel, and Chuck Collard of the Conoco bulk plant had not decided on future plans but Dudley and Tomlinson of the Center Grocers had purchased land in Frisco. Helen Christian of Blue River Inn did not wish to comment for the newspaper article. The Arapahoe Café and Holiday House Motel expected to move to the new town at some future date.

Of the Dillon businesses, therefore, two planned to go out of business, two expected to close for the winter, one would move out of state, four would go to Frisco, and eight to Silverthorne while eight hoped to continue in Dillon for the time being. Six of them had not decided on a future location and two had tentative plans for new Dillon.

The reporter summarized the feelings of many: "Dillon residents, when they sold their properties to the Denver Water Board, had been promised that a new town would be provided for them, from which the Water Board would not profit, so that the historic old town could continue its existence in a nearby location. But the Water Board, no matter what it claimed, by forcing action before the new town was ready for occupancy, scattered the people of Dillon in every direction so that the new town would bear little resemblance to its predecessor." (72) **(Figures 40, 41, 42, 43), (Figures 44, 45, 46), (Figures 47–65)**

Figures 40, 41, 42, 43. The Demolition of Dillon Drug and Kremmling's Store on Main Street, Dillon. (Courtesy Board of Water Commissioners, City and County of Denver)

Figure 41.

Figure 42.

Figure 43.

Figures 44, 45, 46. *The Demolition of the Lucky Horseshoe Bar and Café. Lake Hill and the west abutment rise behind the workers. (Courtesy Board of Water Commissioners, City and County of Denver)*

Figure 45.

Figure 46.

Figure 47. Karl's Conoco at Corner of Routes 6 and 9. Behind the buildings is Lake Hill. (Courtesy Board of Water Commissioners, City and County of Denver)

Figure 48. Bulldozed Buildings. Some buildings were destroyed on site and not moved to the new town.

Figure 49. The Remains of Dillon Garage. The overlook at the west abutment would be built above the peaked roof of the Dillon garage. (Courtesy Board of Water Commissioners, City and County of Denver)

Figure 50. Heart of the Rockies Cabins. During the summer season, many tourists stayed at the Heart of the Rockies Cabins, built in the early 1930s. (Courtesy Board of Water Commissioners, City and County of Denver)

Figure 51. Burning what can't be Moved. Fires burned for days, consuming the remains of anything that not could be hauled away. (Courtesy Board of Water Commissioners, City and County of Denver)

Figure 52. Tenderfoot Ranch. The Byers family sold almost all of their Tenderfoot Ranch to the Water Board. Part of the ranch became the Water Board's permanent headquarters in Summit County. (Courtesy Board of Water Commissioners, City and County of Denver)

Figure 53.

Figure 54. Wildwood Lodge on the Way to Silverthorne. One of the last to leave the reservoir site, the Wildwood Lodge served meals to the workers until the very end. (Courtesy Board of Water Commissioners, City and County of Denver)

Figure 55. The Final Two. Dillon Community Church and the 1910 school sitting side by side remained until the end of the school year. Students who could not fit into the classrooms of the 1910 school attended classes in the church. (Courtesy Board of Water Commissioners, City and County of Denver)

Figure 56. Packed Classrooms. One teacher in the 1910 school taught 70 seventh graders in his math class; he tended to the needs of 48 eighth graders in his science class. (Courtesy Summit Historical Society)

Figure 57. Demolition of the 1910 School. Because of its construction and age, the building could not be moved to a new site. (Courtesy Board of Water Commissioners, City and County of Denver)

Figure 58. Ready to Go. The Community Church moved slowly to its new site on LaBonte Street in 1962. *(Courtesy Board of Water Commissioners, City and County of Denver)*

Figure 59. Missing Annex. Before the building could be moved, workers removed the annex where Sunday school classes had been taught. *(Courtesy Board of Water Commissioners, City and County of Denver)*

Figure 60. The Inlet Structure near the West Abutment. The church passed the inlet structure on the way to new Dillon. (Courtesy Frisco Historic Park & Museum and Polhemus Family Collection)

Figure 61. Along the Haulage Road. Trucks pulled the Community Church along the haulage road. Notice the "fins" of the Glory Hole in the background. (Courtesy Board of Water Commissioners, City and County of Denver)

Figure 62. Overhead Wires. At one point, overhead wires stopped progress along the haulage road. The wires could not be lifted; the cupola could not be removed. An excavator lowered the roadbed so that the church could continue on to new Dillon. (Courtesy Frisco Historic Park & Museum and Polhemus Family Collection)

Figure 63. The East Abutment. Notice the road cut to the far left of the steeple that will become the road over Dillon Dam and the grout cap almost directly below it.. (Courtesy Board of Water Commissioners, City and County of Denver)

Figure 64. Sunday School Annex. The annex arrived separately at the new site on LaBonte Street. (Courtesy Frisco Historic Park & Museum and Polhemus Family Collection)

Figure 65. Ready for Services. Dillon Community Church held its first services in its new home on July 1, 1962. The Ladies Aid Society of the Church with the financial help of Mr. Potashnick assured that the new town would have its church. (Courtesy Summit Historical Society)

New Dillon

MANY SITES HAD BEEN CONSIDERED for the new town. Some people showed interest in building new Dillon on the highway north of the dam, the location now occupied by Silverthorne. (73) But after much hesitation because a decision on the relocation of Highway 6 had not been made, the town board chose the ridge north of old Dillon and south of Straight Creek. To learn whether town residents approved of the choice, the town board circulated petitions and placed some in businesses. Signers agreed that the mayor and town board should have the power to relocate Dillon to that spot. (74)

Because the town of Dillon did not have the necessary financial resources to establish the new town, the Denver Water Board advanced Dillon $10,000 for planning and allied services. It was not a loan. Rather, the amount would be credited against any payment the Water Board might be required to make to the town or county for land, roads, buildings, or other property. (75)

After being awarded the contract for designing the new town, the firm of Trafton Bean interviewed residents to learn what they thought the town's form and function should be.

Some preferred a town with a business district in a strip along U.S. Route 6. Others preferred a town with a central shopping area surrounded by residences. Still others felt the town should function as a highway stop much as old Dillon had. Many favored a year-round resort town. One resident, writing in the *Summit County Journal* asked, "Do we want a Stringtown, an old fashioned hick town, the same as our old Dillon? Or do we want a new modern, up-to-date planned town? Why go backwards? This is the 'chance of a lifetime.'" She preferred a circular plan with shopping area in the center rather than a stringtown development. (76)

A state highway department ruling that would not allow the business district of the new town to be put in a strip along Route 6 actually decided the town's arrangement (strip vs. circle development). Only one access road would be allowed into the new town from Route 6 as it was intended to be a limited-access highway.

Thus, when Trafton Bean unveiled a scale model of new Dillon to the town board on December 10, 1958, they and others saw a compact business district located along the one and only access road to Route 6. Two major residential areas, protected by topography from winter winds, would be located on either side of the Central Business District. Recreational and tourist facilities would be along the waterfront. In answer to questions as to why the businesses could not line the highway, Mr. Bean explained that "State Highway Department policy did not allow that, and that it will allow only one access road to the business area." Therefore, the current plan took best advantage of that. He added that with only one access road to the business district, traffic would flow around the business district rather than through it. The minimum of roads would reduce the maintenance and snow removal problem. (77) **(Figures 66, 67, 68, 69, 70)**

Maps of the preliminary plan had been posted in the Dillon Market windows for several weeks before the meeting.

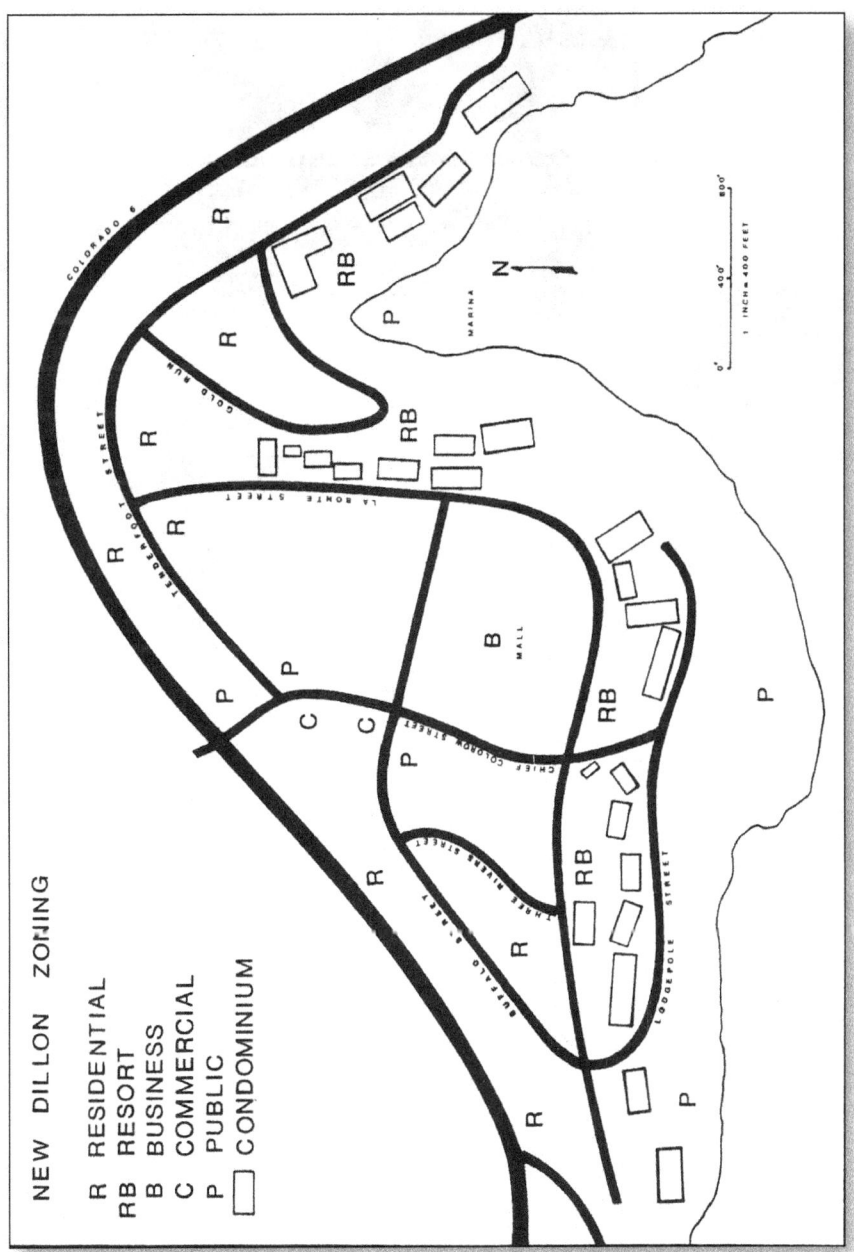

Figure 66. New Dillon. Designed by Mr. Trafton Bean and his associates, the town had designated areas for residential, resort, and commercial construction. (Adaptation drawn by Author)

Figure 67. Dillon Town Hall. The social center of the old town, the Town Hall housed the post office, an apartment for the marshal, and a space for the town's fire engine. (Courtesy Frisco Historic Park & Museum and Polhemus Family Collection)

Although the plan received general approval, discussion centered on whether the new town should focus on year-round resort business or aim to attract the one-stop highway trade that now made up its present livelihood.

Some favored the resort business and its more even distribution of income over the year, rather than the summer surge followed by the slow winter. The town's location on Highway 6 and potentially the proposed interstate highway offered a variety of promotional opportunities.

Not one to give up easily, Virgil Cox, who operated the Dillon Inn, presented a map suggesting that a location along U.S. Highway 6 north of the town site offered a more advantageous site for the businesses. W.A. Doyle of Colorado Springs, owner of the Holiday House Motel, agreed. The group asked Trafton Bean to develop an alternate plan using

Figure 68. 1882 Opera House. Built in Frisco by William Graff, who moved it to Dillon in 1887, the Opera House became the meeting place for several Rebekah lodges in town. (Courtesy Frisco Historic Park & Museum and Polhemus Family Collection)

a commercial street, rather than the shopping center idea, while the town board would proceed with plans for the town at the agreed-upon location—despite the fact that the additional planning could delay the completion of the new town for about a year. (78)

Once they saw the scale model of the town, Dillonites began to worry that the Water Board would not actually help install the water and sewer systems in the new town as promised. This, combined with the fact that the Water Board still did not own vital land within the reservoir site, caused many residents and business owners to delay making definite plans for the future. (79)

Wanting to begin developing the new town site and thus showing the doubters that there indeed would be a new Dillon, town leaders asked the Denver Water Board for a loan

Figure 69. Blue River Lodge. The Blue River Lodge met in the former 1882 Frisco Opera House until the Lodge ceased to exist. (Courtesy Frisco Historic Park & Museum and Polhemus Family Collection)

of $440,000 to begin construction on the water and sewer systems for the new town. The Water Board refused this request stating that under its operating policies, the Board could not use Denver money for real estate development. (80) But when final decisions on highway relocation and financial support for the high dam had been made, other matters fell into place. Dillon would receive $450,000 for sewer and water systems. Not a gift and not a loan, the money would be repaid through land sales and service fees. (81)

Figure 70. The Ray and Marge Hill House. Movers experienced great difficulty pulling the very heavy log house up the hill to its new home on LaBonte Street. (Courtesy Frisco Historic Park & Museum and Polhemus Family Collection)

The final agreement between the town of Dillon and the Water Board appeared in the newspaper on May 27, 1960 (page 1). It specified that not before April 1, 1961, would the present site have to be vacated. By that date, all of Dillon's property in the reservoir site would be sold to the Water Board for $12,000. (Later the *Summit County Journal* reported that the amount was raised to $12,500.) (82) In turn, the Water Board would convey to Dillon approximately 142 acres for the new town site at a cost of $35 per acre. This cost plus the money used to build the water and sewer systems would be repaid by the sale of residential and business lots in the new town. The plan called for a total of 112 residential sites. (83) Of the selling price, 85 percent would go to repay the debt. Each quarter, a minimum of 50 percent of the gross revenue from the operation of the water and sewer systems would go to the Denver Water Board until the amount was satisfied in full. All county lands within the reservoir site, including highways and roads, would become the property of the Denver Water Board in exchange for an equal amount of Water Board land outside of the reservoir site and new town site. The Water Board would also pay for the cost of relocating any buildings that had to be moved from county land within the reservoir site. (84)

Half-acre lots in new Dillon sold for $2,000 up to $3,250. In some cases, these prices exceeded what the owners had been given for their land when they sold to the Water Board. One-acre commercial lots cost between $6,500 and $10,000. The cheapest lots were along the lake; people still feared that the engineers miscalculated and these parcels would be under water. (85)

Until Dillon repaid its debt to the Water Board, the Water Board had the right to sell land within the new town site. But once the debt had been satisfied, any unsold land would revert to new Dillon. For purposes of selling the commercial and residential sites, the town board became the Dillon Improvement Corporation and its articles of incorporation appeared in the *Summit County Journal* on June 3, 1960. (86)

The town enforced strict building codes. Structures could be no higher than 30 feet to avoid blocking views of the lake. Contractors should use materials such as wood, stone, and glass in construction; trees should be cut sparingly; signs must be subdued; and the codes permitted no shacks, dead autos, or house trailers. (87)

The differences between the two towns are striking. Old Dillon was a western town—a watering hole for the cowboys; new Dillon is a resort town—anytown, USA. Old Dillon was home to some families who had lived in the valley for generations; new Dillon drew second home owners whose needs, desires, and wants often differ from those of year-round residents. Old Dillon served as a crossroads town. Major routes in the county crossed in Dillon; the Denver, South Park & Pacific and the Denver, Rio Grande railroads both served the town. New Dillon had only one entrance to town; no east or west entrances—only the main entrance on Chief Colorow Drive (now Lake Dillon Drive) and no signage for businesses could be placed at the entrance because of state policy. Old Dillon was an agricultural town; new Dillon is a resort town. Old Dillon had a long and proud history; new Dillon had no history. Old Dillon knew its place in the county—transportation hub—agricultural town—old western town. New Dillon had to establish itself and determine what it wanted to be "when it grew up."

Dillon Cemetery

PERHAPS THE MOST UNIQUE ITEM MOVED was Dillon Cemetery, the only cemetery patented under federal law. President McKinley signed the patent on June 30, 1901. Since the old cemetery would be inaccessible once the reservoir filled, the cemetery, established in 1885 on 53½ acres of land donated by Chauncey Warren, had to be moved. The Water Board agreed to pay the Western Vault Company of Holyoke, Colorado, to disinter and reinter the more than 300 graves at the new 38-acre-site located east of the new town along U.S. Route 6. *(Figures 71, 72)*

"Plans are now complete for the moving of the old Dillon cemetery to the new location. The graves will be relocated in an arrangement as nearly similar to that of the old locations as is possible.

There are a total of 327 graves to be moved, 325 of which will go to the new location. Two will be taken to Buena Vista for interment.

The Dillon cemetery is older than the town. The oldest readable headstone bears the date 1879, and 133 are buried there before 1911. In the period between 1911 and 1930 there were 78 burials, from 1931 to 1940 there were

Figure 71. New Cemetery. Because the old cemetery would be inaccessible, over 300 graves had to be moved to the new site along Route 6, east of town. (Courtesy Board of Water Commissioners, City and County of Denver)

Figure 72. New Grave Site. Historic names grace the headstones in the new cemetery: Ballif, Cluskey, Lindstrom, Lund, Mumford, Warren, Roby, Giberson, and Laskey to name a few. (Courtesy Board of Water Commissioners, City and County of Denver)

54, from 1941 through 1950 there were a total of 46, 14 were buried in 1950 to 1960 and two from then to date (1962). Of the 327 graves 37 are infants, 19 are children, and 271 are adults.

The tombstones, markers, fences, etc., will also be moved and placed in proper position. The removal will be done in a dignified and private manner. Suitable containers made of Ponderosa pine or Red cedar will be provided.

Among the historic names are Ballif, Christensen, Cluskey, Lindstrom, Lund, Markey, Marshall, McGee, Ashlock, Mumford, Warren, Roby, Giberson, Altland, Philips, Jobe, Laskey, Wiley, Wehrly, Tubbs, and many others who left their mark on the history of the county.

The contract has been awarded to the Western Vault Company of Holyoke, Colorado. Their bid was $21,980.

They have agreed to the many and detailed requirements set forth by the Town of Dillon and the Denver Water Board." (88)

Unknowns lie in the new cemetery—but these unknowns are not truly unknown. Rather the designation indicates individuals whose relatives could not be found at the time of the move. A letter "D" within a circle on a headstone designates a grave moved from the old cemetery to the new.

Swan Mountain Road

MANY COUNTY RESIDENTS EXPRESSED CONCERN about the geographical isolation that the reservoir would create for those living far from the center of the county. Residents felt that there should be roads on all sides of the reservoir to prevent this isolation. In particular, the need for a road on Swan Mountain created the greatest concern. Without it, each trip to Breckenridge from Montezuma required many additional miles. Children would travel extra miles each day going to and from school. As a reporter for the *Summit County Journal* noted: "If the Denver Water Board doesn't build the road, Summit County would have to do it later." (89)

Hoping to escape the extra costs of building a road on Swan Mountain, the Water Board tried to link the money for Dillon's water and sewer systems to road construction. The Board said it would advance the funds to Dillon if Summit County would give up its demand for a road on the eastern side of the reservoir as well as its rights of reversion to state highways that must be abandoned. Few in the county approved of this proposal. (90)

In another attempt to avoid building the road, the Water Board agreed to obtain or pay costs for obtaining new rights-of-way for those parts of U.S. Route 6 and Colorado Route 9 displaced by the project. In return, the county would not require the Denver Water Board to construct a road over Swan Mountain. (91)

At a time when nerves were already stressed, insensitive statements made by individuals working for the Water Board created additional anxiety and concern. During a discussion about the road over Swan Mountain, according to a *Summit County Journal* reporter, a Water Board lawyer allegedly stated that Montezuma people go to Breckenridge once every five years and do not want a road on Swan Mountain; that no people lived in the Montezuma end of the county; and that the Water Board would not be forced to build the road over Swan Mountain. (92)

The chief reason appeared to be that the Water Board did not have funds to build the road. County residents felt that they needed roads on all sides of the reservoir to replace those taken away and funding was not the county's worry. One suggested that if Denver "must have the dam in Summit County, then Denver must pay for adequate replacement of the roads they are taking." (93)

In the final agreement, Swan Mountain Road would be built and paid for by the Water Board. (94)

Completing the Dam and Water Storage

BY THE END OF MAY, 1963, only 32 feet remained to be added to the crest of the dam. The expected completion date was July 15, 1963. Taking advantage of the snow-free season, 170 men worked two ten-hour shifts per day. (95) Because of this, they topped out the dam on July 1, two weeks ahead of schedule.

Water storage using water from the Blue River began September 3, 1963, at 9:00 a.m. (96) Crews shut off the flow of the Blue River by forming a small coffer dam with bulldozers. *(Figure 73)*

Officials expected the reservoir to fill in three or four years. In the first week of water storage, the water rose six inches per day. This rate, of course, would decrease as the water surface increased. (97)

Two of the last three buildings in the old town, the Antlers Café and Bar, which was taken to Frisco, and the Wildwood Lodge, which went to Silverthorne, were moved about three weeks before water storage began. (98) The Arapahoe Café, now in new Dillon, was also among the last to be moved. As the water rose, one house, that of Basil French, remained on

Figure 73. Steadily Rising. Water storage began on September 3, 1963 at 9:00 a.m. Water first flowed over the Glory Hole on August 3, 1965, at 3:00 a.m. (Courtesy Board of Water Commissioners, City and County of Denver)

site. The lone residents of old Dillon had not yet settled with the Water Board. (99)

Water from the Snake River and from Straight Creek satisfied water rights downstream from Dillon. (100) But because of this, Green Mountain Reservoir did not receive its usual amount of water. (101)

When the federal government, which controlled Green Mountain Reservoir as part of the Big Thompson Project, and western slope irrigation districts realized the impact water storage in Dillon Reservoir had on their interests, they asked the United States District Court to halt the filling of the reservoir. In response, the Water Board guaranteed that no water would be held in Dillon Reservoir until Green Mountain Reservoir reached capacity the next year. The Water Board also offered to allow water to be taken from the Dillon site if Green Mountain did not fill. (102) Because this answer did not satisfy the plaintiffs, the suit was not withdrawn. Six months later, the Court ruled that the Water Board could divert water for the reservoir as long as it did not violate the right of the federal government to fill Green Mountain Reservoir. (103)

Several months earlier as winter approached, low water levels endangered fish in the Blue River between the two reservoirs. An amount of water equal to that entering the Dillon Reservoir (150 acre-feet per day) was released to prevent freezing and loss of fish. (104)

Some Final Statistics

EXCEPT FOR NORMAL WINTER SHUTDOWN, no weather delays slowed construction. (105) Unfortunately, two workers died. On August 17, 1960, a dump truck skidded off a slippery approach road, tipped over, and landed on the driver, who had been thrown from the vehicle. The second death occurred May 21, 1963, in Borrow Area #1 when an empty 17-cubic-yard bottom-dump Euclid truck collided with a loaded truck of the same size. (106)

The actual completion date of the dam was December 17, 1963, although the final inspection and acceptance of both the dam and reservoir occurred January 10, 1964. (107)

With a shoreline of 24.5 miles and a 3,300 acre surface, the reservoir's maximum elevation is 9,025 feet. The top of the Glory Hole spillway has an elevation of 9,017 feet. Total storage capacity measures 257,304 acre-feet or about 85.5 billion gallons. Of that, 249,556 acre-feet is live storage; the rest is dead storage for sediment carried into the reservoir. (108) At the time of completion, Denver owned the water rights to 252,678 acre-feet of water in the reservoir. The dam is 231 feet tall, 5888 feet long, and required 12 million

Figure 74. Completely Filled. With a shoreline of 24.5 miles, the reservoir holds 85.5 billion gallons or approximately 257,304 acre-feet of water when full. (Photograph by Author)

tons of fill. The crest width measures 32 feet while the base width extends 2,700 feet. (109)

The *Summit County Journal* reported that the Water Board spent a total of $1.8 million for property in the reservoir site (1961 dollars), while the Water Board calculated the total cost of the dam and reservoir at $19,423,297.70 (1961 dollars). (110, 111) **(Figures 74, 75)**

The dam undergoes an annual safety inspection. The *Summit Daily News* on May 31, 2001, reported that the dam seeps about 50 gallons of water per minute. The seepage, near Anemone Trail, is collected and enters the sewer system. A full scale breach exercise occurs every ten to 15 years. In the very remote possibility of a dam failure, an 80-foot-tall wall of water would bury Silverthorne, race toward Green Mountain Reservoir, taking three hours to get to the reservoir, and then overtop Green Mountain Dam.

Figure 75. Dillon Dam and Reservoir. The Summit County Journal reported that the Water Board spent a total of $1.8 million for property in the reservoir site (1961 dollars), while calculating the total cost of the dam and reservoir to be $19,432,297.70 (1961 dollars). (Photograph by Author)

In 2010, the Water Board reported that about 54,000 acre-feet of water goes through the Roberts Tunnel each year. The original low dam that the Water Board had envisioned would have impounded only 52,000 acre feet of water.

1931–1964

Time Line

July, 1931: Core samples taken along tunnel route

November 14, 1942: Location plans filed

1945: Study conducted by Ogden Tweto

February 16, 1946: Approval for construction obtained

June 24, 1946: Work on tunnel resumes

November 3, 1955: Denver Water representatives meet with residents at Wildwood Lodge

December 10, 1958: Trafton Bean and Associates unveil model of new Dillon

January, 1959: Denver Water representatives still buying needed land for reservoir

October 27, 1959: General contract awarded to Potashnick Corporation, Inc.

December, 1959: Eviction letters mailed specifying April 1, 1961, date

January, 1960: Condemnation proceedings begin

January 2, 1960: Two drilling teams meet in western portion of tunnel

February 24, 1960: Two drilling teams meet in eastern portion of tunnel

April 22, 1960: Clearing of vegetation begins

May 10, 1960: Vote taken authorizing high dam

May 27, 1960: Agreement reached between Denver Water Board and Dillon relative to evacuation of town and roadways, purchase prices for infrastructure and land, the $450,000 advance for water and sewer and repayment rules, etc.

June 4, 1960: Modified contract arranged with Potashnick Corporation, Inc.

August 5, 1960: Plans revealed to purchase buffer strip

April 1, 1961: First eviction date announced—later revised

June 1, 1961: School term finished

June 9, 1961: Some necessary land for reservoir still not purchased

August 1, 1961: Final eviction date set for everyone

July 21, 1962: 2-4-D sprayed to kill vegetation

July 1, 1963: Dam completed

August, 1963: Last buildings vacated the town

September 3, 1963: Water storage begins

November, 1963: French house remains on site

December 17, 1963: Entire project completed

January 10, 1964: Final inspection conducted

July 17, 1964: First water moves through tunnel

How Dillon Received Its Name

ON JULY 26, 1881, THE DILLON MINING COMPANY, led by Harper M. Orahood, Peter Halbert Sayre (Hal Sayr), and others of Denver, patented a town site of 320 acres stretching mainly to the northeast of the Snake River. This was the first town site. Company members included savvy businessmen who expected the railroad to lay its tracks over one of the mountain passes from Denver, along the Snake River, and on to Leadville. They knew that if they laid out their town with surveyed lots, the value of the land would increase in anticipation of the railroad's arrival. These real estate speculators hoped to make their fortunes by selling lots in their town. *(Figure 76)*

Sayre (Sayr), one of the company leaders, served as a trustee of the Denver, Georgetown, Utah Railway Company in 1872. As a trustee, he knew the plans proposed by various companies for building into the mountains. Both the Georgetown, Leadville & San Juan Railroad Company and the Georgetown, Breckenridge & Leadville Railway Company intended to build a route over the Continental Divide and along the Snake River.

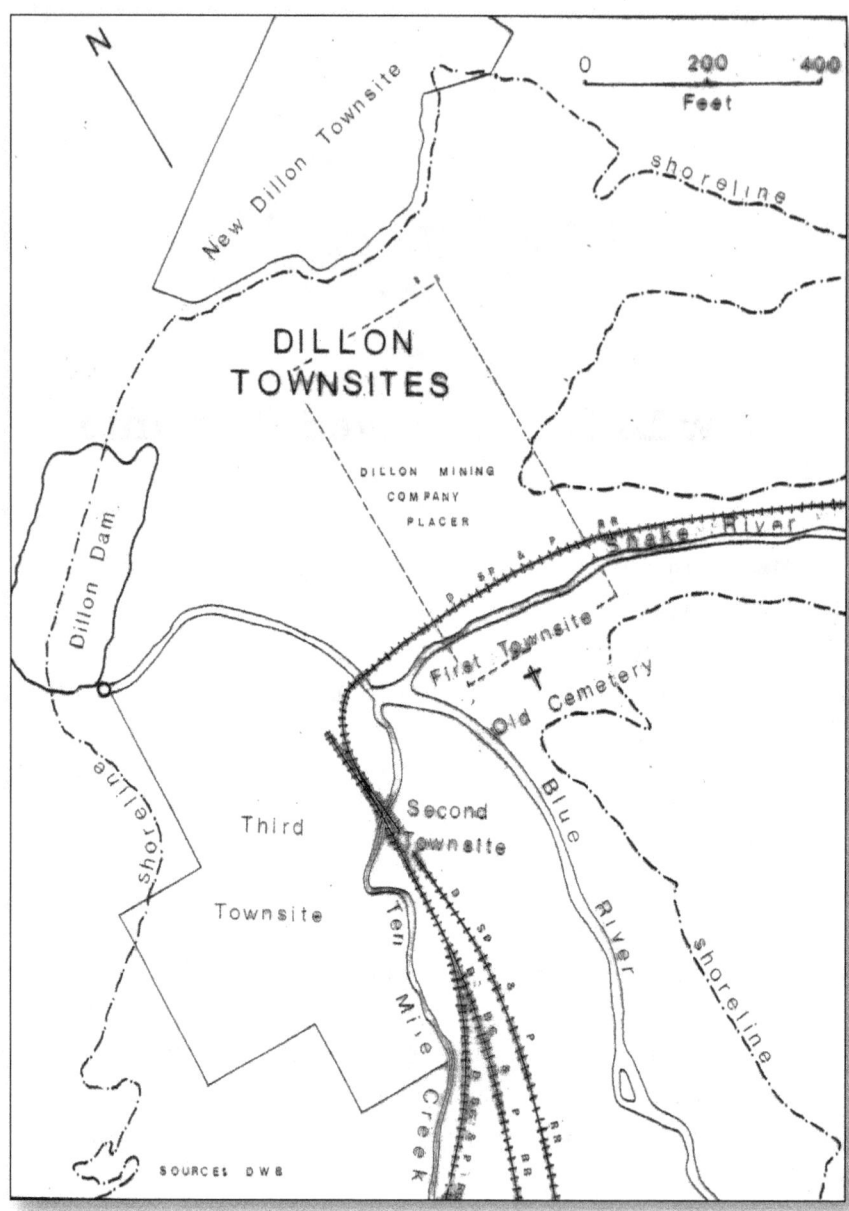

Figure 76. Four Dillon Town Sites. (Map drawn by Author)

How Dillon Received Its Name

Wanting to make sure that the railroad crossed their town site, the leaders of the company named their town for Sidney Dillon, called an entrepreneur and an incorporator for the Union Pacific Railroad. In March, 1881, Dillon became the president of the Union Pacific. In 1883-1884, Dillon served as president of the Colorado Central, part of the Union Pacific system, as well as a director of both the Georgetown, Leadville & San Juan Railroad and the Georgetown, Breckenridge & Leadville Railway Company. Four months after Dillon became president of the Union Pacific, the Dillon Mining Company laid out its town named for him.

The town's first two moves reinforce the importance of the railroad to the town and the fact that the town bore the name of this important railroad executive. The first move, to a spot between the Blue River and Ten Mile Creek, incorporated the tracks of the Denver & Rio Grande that arrived in November, 1882; the second move, a month later to a site west of the Ten Mile Creek, assured that the town now included the newly arrived tracks of the Denver, South Park & Pacific as well as those of the Denver & Rio Grande.

The most recent move occurred as a result of the flooding of the third town site. Thus, Dillon is on its fourth town site; it has moved three times.

New Dillon Walking Tour

WHEN THE RESIDENTS OF OLD DILLON realized that the dam and reservoir would become a reality, they decided not to let their town die. With the financial help of the Water Board, they chose Mr. Trafton Bean to design their new town on part of the Tenderfoot Ranch, which had been sold by Mr. and Mrs. Frank Byers to the Water Board.

When the time came to vacate the old town site, some owners moved their buildings to Breckenridge, Frisco, or the recently established town of Silverthorne; other structures became the first to line the streets of new Dillon.

On this walking tour, you'll not only see those buildings moved from the old town to the new; you will see some that were moved from sites elsewhere in the county. For some, it was not the first move. A few of these buildings followed the town as it moved to assure its place along the tracks of the county's two narrow-gauge railroads.

 Begin at the Summit Historical Society's Schoolhouse Museum at 403 LaBonte.

403 LaBonte: This building, constructed in 1883, housed the first school in old Dillon. When the town built a new school in 1910, this became the community church. Potashnick Construction, Inc., the general contractor for the dam, donated funds to move the building to its present site. When the congregation completed its new church next door, the old church became the Summit Historical Society's Schoolhouse Museum. Inside is a schoolroom furnished with artifacts from a variety of one-room schools in the county.

Behind the Schoolhouse Museum is the Myers Cabin, built in 1885 as part of the Delker ranch. The hand-hewn structure sat near the banks of the Snake River, where the base of the gondola is now located at Keystone Resort. Charles Delker was employed as a mine manager and millwright in Leadville and at the Colorado Toledo Mining Company in Montezuma. His wife, Marie, known as Nurse Delker, is buried in Dillon Cemetery. In the early 1920s, his heirs sold the ranch to Dimp Myers, a mining engineer associated with many mining enterprises in the county, including the King Solomon Mining Syndicate in Frisco. Following Dimp's death in 1954, Lula lived in the cabin until 1966. The Historical Society moved the cabin to its present site in 1976.

The Honeymoon Cabin, supposedly built by a man for his bride, has the Scandinavian style of construction with peeled logs, diminishing in size with height. The very small door preserves heat—it does not indicate a short owner. A skilled lumberman could build a cabin similar to this without assistance from others. Cement caulking fills the spaces between the logs on the outside while slender aspen boughs serve the same purpose on the inside.

 Leave the Schoolhouse Museum and turn north (left) on LaBonte.

415 LaBonte: Ray and Marge Hill built this log house in 1947, one of the first to be moved to new Dillon. North of the driveway, behind the larger house, stands a cabin that was also part of old Dillon. In 1961, the Hills purchased two lots in the new town site. On moving day, the heavy house, sitting on a flatbed truck, quickly became stuck. The house sat in the same spot for at least two weeks before arriving in its new Dillon location.

435 LaBonte: This house, not native to Dillon, originally stood on a ranch on the lower Blue River. Owned by Jim Smith and his son, Dick, it was moved first to old Dillon and then to the new town site.

205 Tenderfoot: John Younger and his wife, Stoney, who died suddenly in 1982, moved the front central portion of this house from the old town in 1961. The house has floors of wide pine boards. At one time, Emma Baliff, wife of Tip, mother of Anna, and grandmother of Viola and Henry Emore, lived here. The wooden board and batten garage at the rear of the property arrived in new Dillon on the back of a dump truck.

John Younger, who worked on the dam and tunnel, served as Dillon town clerk and building inspector, and as a member of the town council for over ten years.

A still-visible road bed cuts through the Younger property. The road ran from Straight Creek, across the property, through what is now the Lake Dillon Amphitheatre, and on into old Dillon. A road branching off from the main road led to the Byers ranch. Another branch led toward the twin bores of the Eisenhower/Johnson tunnels. Much of this branch of the road was at one time a corduroy road, so named for the logs placed in the roadway to keep wagon wheels from becoming mired in the mud.

175 Tenderfoot: Born in 1919, Sena Valaer, daughter of Oscar and Ambure Otterson, lived here for many years. The Ottersons were a Breckenridge family. Sena's husband, John, plowed the streets of Dillon in winter. Residents were happy to see John at the wheel of the snowplow because he would be sure not to plow snow back into their driveways as he cleared the streets. Behind the garage sits the building that housed the Home Café in old Dillon. The front of the structure faces Highway 6 because the Valaers intended to open it for business but Highway Department policy would not allow businesses facing what they planned as a limited-access highway.

159 Tenderfoot: Set at an angle, the central part of this building served as part of the Forest Service complex at Dickey Ranger Station, located at the southern end of the reservoir near Farmer's Corner. Frank Byers, who owned the ranch on which the new town is located, moved it to old Dillon. Mike Uitich subsequently moved it to new Dillon.

156 Tenderfoot: Standing on Hamilton Street in old Dillon, this building housed the *Blue Valley Times,* founded by John Leuthold. A native of Zurich, Switzerland, and an ordained minister, he learned the printing trade from a tramp printer while working as a runner for railroad hotels and boarding houses in Kansas. He published several books on printing that were in their time regarded as standards in the field. After the death of his wife and son, he spent the later years of his life at the International Typographical Union Home in Colorado Springs where he died at the age of 92. He is buried in Valley Brook Cemetery in Breckenridge with his wife and son. After the move from old Dillon, the newspaper office, still clad in its original siding, became a residence.

 Cross Lake Dillon Drive and walk behind the service station.

134 Lake Dillon Drive: Many residents of Summit County considered membership in fraternal organizations and lodges a means of overcoming isolation. Membership provided a link to a national and even international community. Both the Rebekahs and Odd Fellows, often called the poor man's Masons, thrived in Summit County. The Three River Rebekah

Lodge #135 met in this building on Hamilton Street. Sitting one block north of the Hamilton Hotel and next to the Younger house in old Dillon, it was originally the 1882 Graff Opera House.

The Graff Opera House opened its doors in Frisco on February 29, 1882, with people from around the county enjoying "A Happy Pair," a comedietta in one act, and "Turn Him Out," a farce. The play, "All that Glitters is not Gold," a comic drama, opened in October of that year. Admission cost 50 cents. Many traveling groups performed in the Opera House. The Frisco Dramatic Society staged numerous productions.

William Graff, from a wealthy stove and iron manufacturing family in Pittsburgh, built the opera house in 1882. Previously he had been active in mining in the Kokomo area. He managed an unsuccessful smelter, owned an interest in the Queen of the West mine, and directed the Kokomo Brass Band, earning the name Professor Graff. He served as mayor of Frisco in the 1880s. The Opera House remained in Frisco until 1887 when horses pulled the structure on skids to Dillon. He resided in Dillon until his death in 1907 at the age of 81.

The Rebekahs of Blue River Lodge #45 held their meetings in this building. This lodge later consolidated with Dillon Lodge #164, which had formed in 1910, becoming Blue River Lodge #49. Three Rivers Rebekah Lodge #135, which dates to 1912, met in the building until membership dropped and the lodge ceased to exist.

176 Lake Dillon Drive: In front of the lodge building stands the old town hall built in 1899. The social center of old Dillon, residents enjoyed dances and special events held in the building. In 1954 the post office moved into the front of the building replacing the town's theatre. Town Marshal, H.H. Ross, lived in the apartment at the rear. The town's fire engine occupied the middle of the building. The fire engine doubled as the first meeting room for the Dillon town board. Members of the board sat on the fire engine while discussing town business. Many buildings had to be moved from old Dillon at a moment's notice. A story is told that Susie Thompson, the postal clerk, continued sorting the mail as the building moved to its new site.

 Continue south on Lake Dillon Drive. Turn west (right) on West Buffalo.

300 West Buffalo: Chuck Collard, who came to Dillon from the town of Como, where he was a member of a prominent ranching and mining family, occupied this house. At the time of his death, he held the position of Summit County Commissioner.

304 West Buffalo: Henry Emore lived here in one of the first houses to be moved to new Dillon. He and his sister, Viola, who lived on Tenderfoot, were the third of five generations to call Summit County home. His grandfather, Tip Ballif, a native of Bern, Switzerland, and a trained veterinarian, first came to Summit County to drive for the High Line Stage Coach Company, which carried passengers from Georgetown to Leadville. He opened a blacksmith shop in Frisco, the town's first, and later moved the business to Dillon after the railroad arrived in 1882. He also had a blacksmith shop in Keystone.

Tip served four terms as Summit County sheriff after several years as deputy sheriff. He wore the deputy's badge in 1898 when Pug Ryan (namesake of the Dillon restaurant) was arrested for the deaths of two deputies in a shoot-out after the robbery of the Denver Hotel in Breckenridge.

Return to Three Rivers and turn south (right).

109 Three Rivers: The Dillon town board moved and refurbished the one-story part of this home to be the living quarters for Bessie Blundell, the widow of Postmaster Ira Blundell and daughter of Chauncey Warren. The grandson of Swedish immigrants, Ira Blundell married Bessie Warren in 1908, and remained a resident of the town until his death.

Chauncey Warren ran a stage coach station at the foot of Loveland Pass that served the High Line Stage Coach route from Georgetown to Leadville from 1879 to 1882. Settling in Dillon, he built the first hotel, operated a stable, served as mayor, and donated 53 acres for the town's cemetery. His son, Brad, at age 17, was one of the eight drivers on the Georgetown to Leadville route.

 Turn east (left) on LaBonte to Lake Dillon Drive.

26 Lake Dillon Drive: One of the last buildings to be moved from the old town site was the Arapahoe Café. Owned by Faye and Lenore Bryant, it remained on its site to serve the construction workers. Later, after it was moved to new Dillon, Mike Uitich operated the restaurant—for a short time after the move, it was the only restaurant in new Dillon. It reopened for business on May 13, 1962. The Arapahoe Café has earned a reputation for the best BBQ in Summit County.

References

1. Board of Water Commissioners, City and County of Denver. *The Blue River Diversion Project.* July 1964, 1-3.
2. Board of Water Commissioners, City and County of Denver. *The Blue River Diversion Project.* July 1964, 4.
3. Board of Water Commissioners, City and County of Denver. *The Blue River Diversion Project.* July 1964, 4.
4. Board of Water Commissioners, City and County of Denver. *Final Construction Report for the Dillon Dam and Appurtenances.* March 1964, 2; Board of Water Commissioners, City and County of Denver. *The Blue River Diversion Project.* July 1964, 13.
5. Board of Water Commissioners, City and County of Denver. *The Blue River Diversion Project.* July 1964, 13.
6. *Summit County Journal,* 7 October 1955, 1.
7. Board of Water Commissioners, City and County of Denver. *The Blue River Diversion Project.* July 1964, 5. Denver Water Board Hydropower brochure.
8. Tweto, Ogden et al. *Geology Report on the Dillon, Two Forks Dam Sites, South Platte Canyon.* Washington, D.C.: United States Government Printing Office, 6 February 1945, 1.
9. Tweto, Ogden, et al. *Geology Report on the Dillon, Two Forks Dam Sites, South Platte Canyon.* Washington, D.C.: United States Government Printing Office, 6 February 1945, 2-7.

10. *Summit County Journal,* 13 July 1956, 1.
11. Board of Water Commissioners, City and County of Denver. *Final Construction Report for Dillon Dam and Appurtenances.* March 1964, 2.
12. *Summit County Journal,* 13 May 1960, 1.
13. Board of Water Commissioners, City and County of Denver. *The Blue River Diversion Project.* July 1964, 9.
14. Board of Water Commissioners, City and County of Denver. *The Blue River Diversion Project.* July 1964, 7.
15. Board of Water Commissioners, City and County of Denver. *The Blue River Diversion Project.* July 1964, 6.
16. Board of Water Commissioners, City and County of Denver. *Final Construction Report for Dillon Dam and Appurtenances.* March 1964, 35
17. Board of Water Commissioners, City and County of Denver. *Final Construction Report for Dillon Dam and Appurtenances.* March 1964, 7.
18. Board of Water Commissioners, City and County of Denver. *Final Construction Report for Dillon Dam and Appurtenances.* March 1964, 81.
19. Board of Water Commissioners, City and County of Denver. *Final Construction Report for Dillon Dam and Appurtenances.* March 1964, 109.
20. Board of Water Commissioners, City and County of Denver. *Final Construction Report for Dillon Dam and Appurtenances.* March 1964, 109.
21. Board of Water Commissioners, City and County of Denver. *The Blue River Diversion Project.* July 1964, 8. Dillon Reservoir statistics sheet.
22. *Summit County Journal,* 13 February 1959, 1
23. *Summit County Journal,* 29 July 1960, 1.
24. *Summit County Journal,* 5 August 1960, 1.
25. *Summit County Journal,* 19 August 1960, 1.

26. *Summit County Journal,* 16 September 1960, 1.
27. *Summit County Journal,* 23 September 1960, 1.
28. *Summit County Journal,* 9 June 1961, 1.
29. Board of Water Commissioners, City and County of Denver. *Final Construction Report for Dillon Dam and Appurtenances.* March 1964, 16.
30. Board of Water Commissioners, City and County of Denver. *Final Construction Report for Dillon Dam and Appurtenances.* March 1964, 31.
31. *Summit County Journal,* 27 September 1963, 5.
32. *Summit County Journal,* 25 September 1975, 1.
33. Board of Water Commissioners, City and County of Denver. *The Blue River Diversion Project.* July 1964, 13.
34. Board of Water Commissioners, City and County of Denver. *The Blue River Diversion Project.* July 1964, 11.
35. *Summit County Journal,* 13 July 1956, 1.
36. Robinson, Charles, *et al. General Geology of the Harold D. Roberts Tunnel, Colorado.* United States Geological Survey Professional Paper 831-B. Washington, D.C.: United States Government Printing Office, 1974, 1-2.
37. Robinson, Charles, *et al. General Geology of the Harold D. Roberts Tunnel, Colorado.* United States Geological Survey Professional Paper 831-B. Washington, D.C.: United States Government Printing Office, 1974, 45.
38. Robinson, Charles, *et al. General Geology of the Harold D. Roberts Tunnel, Colorado.* United States Geological Survey Professional Paper 831-B. Washington, D.C.: United States Government Printing Office, 1974, 23.
39. Robinson, Charles, *et al. General Geology of the Harold D. Roberts Tunnel, Colorado.* United States Geological Survey Professional Paper 831-B. Washington, D.C.: United States Government Printing Office, 1974, 1.
40. Board of Water Commissioners, City and County of Denver. *The Blue River Diversion Project.* July 1964, 12.

41. Robinson, Charles, et al. *General Geology of the Harold D. Roberts Tunnel, Colorado.* United States Geological Survey Professional Paper 831-B. Washington, D.C.: United States Government Printing Office, 1974, 23.
42. Board of Water Commissioners, City and County of Denver. *The Blue River Diversion Project.* July 1964, 15.
43. Robinson, Charles, et al. *General Geology of the Harold D. Roberts Tunnel, Colorado.* United States Geological Survey Professional Paper 831-B. Washington, D.C.: United States Government Printing Office, 1974, 24.
44. Board of Water Commissioners, City and County of Denver. The Blue River Diversion Project. July 1964, 17.
45. *Summit County Journal,* 8 July 1960, 1.
46. *Summit County Journal,* 18 September 1959, 1.
47. Board of Water Commissioners, City and County of Denver. *The Blue River Diversion Project.* July 1964, 18.
48. *Summit County Journal,* 8 July 1960, 1.
49. Board of Water Commissioners, City and County of Denver. *The Blue River Diversion Project.* July 1964, 17.
50. *Summit County Journal,* 17 April 1959, 1.
51. Board of Water Commissioners, City and County of Denver. *The Blue River Diversion Project.* July 1964, 18.
52. *Summit County Journal,* 8 January 1960, 1.
53. *Summit County Journal,* 26 February 1960, 1.
54. *Summit County Journal,* 14 December 1956, 1.
55. Board of Water Commissioners, City and County of Denver. *The Blue River Diversion Project.* July 1964, 18.
56. Board of Water Commissioners, City and County of Denver. *The Blue River Diversion Project.* July 1964, 18.
57. Board of Water Commissioners, City and County of Denver. *The Blue River Diversion Project.* July 1964, 21.
58. *Summit County Journal,* 25 March 1960, 1.
59. *Summit County Journal,* 18 May 1956, 1.

60. *Summit County Journal*, 16 December 1955, 1.
61. Board of Water Commissioners, City and County of Denver. *Final Construction Report for Dillon Dam and Appurtenances.* March 1964, 116.
62. *Summit County Journal*, 10 February 1956, 1.
63. *Summit County Journal*, 30 January 1959, 1.
64. *Summit County Journal*, 17 April 1959, 1.
65. *Summit County Journal*, 27 May 1960, 1.
66. *Summit County Journal*, 30 September 1960, 1.
67. *Summit County Journal*, 25 March 1960, 1
68. *Summit County Journal*, 13 October 1961, 1.
69. "Former Dillon Residents. . . . Where are They?" *Summit County Journal*, 13 October, 1961.
70. *Summit County Journal*, 25 March 1960, 1.
71. *Summit County Journal*, 17 May 1963.
72. "Dillon Homes, Businesses to Scatter to Four Winds," *Summit County Journal*, 16 September, 1960.
73. *Summit County Journal*, 25 March 1960, 1.
74. *Summit County Journal*, 7 December 1956, 1.
75. *Summit County Journal*, 6 September 1957, 1.
76. *Summit County Journal*, 4 July 1958, 1.
77. "Only 22 Dillonites attend Meeting on New Dillon," *Summit County Journal*, 19 December, 1958.
78. "Alternate Site for New Town of Dillon Suggested," *Summit County Journal*, 4 July, 1958.
79. *Summit County Journal*, 24 October 1958, 1.
80. *Summit County Journal*, 15 January 1960, 1.
81. *Summit County Journal*, 13 May 1960, 1.
82. *Summit County Journal*, 13 May 1960, 1.
83. *Summit County Journal*, 15 January 1960, 1.
84. *Summit County Journal*, 25 March 1960, 1.

85. Mather, Sandra F. Pritchard. *Southern Summit, A Geographer's Perspective.* Dillon, Colorado: Summit Historical Society, 1984, 42.
86. *Summit County Journal,* 27 May 1960, 1.
87. *Summit County Journal,* 30 September 1960, 3.
88. "Dillon Cemetery Removal to be Started Shortly," *Summit County Journal,* 25 May, 1962.
89. *Summit County Journal,* 12 August 1960, 1.
90. *Summit County Journal,* 25 March 1960, 1.
91. *Summit County Journal,* 25 March 1960, 1.
92. *Summit County Journal,* 18 May 1962, 1.
93. "All Four Roads needed around Dillon Reservoir," *Summit County Journal,* 4 November, 1960.
94. *Summit County Journal,* 4 November 1960, 1.
95. *Summit County Journal,* 31 May 1963, 9.
96. Board of Water Commissioners, City and County of Denver. *The Blue River Diversion Project.* July 1964, 9.
97. *Summit County Journal,* 11 October 1963, 1.
98. *Summit County Journal,* 13 September 1963, 1.
99. *Summit County Journal,* 1 November 1963, 1.
100. *Summit County Journal,* 6 September 1963, 1.
101. *Summit County Journal,* 13 September 1963, 1.
102. *Summit County Journal,* 4 October 1963, 1.
103. *Summit County Journal,* 24 April 1964, 1.
104. *Summit County Journal,* 29 November 1963, 1.
105. Board of Water Commissioners, City and County of Denver. *Final Construction Report for Dillon Dam and Appurtenances.* March 1964, 14.
106. Board of Water Commissioners, City and County of Denver. *The Blue River Diversion Project.* July 1964, 9; Board of Water Commissioners, City and County of Denver. *Final Construction Report for Dillon Dam and Appurtenances.* March 1964, 9.

107. Board of Water Commissioners, City and County of Denver. *Final Construction Report for Dillon Dam and Appurtenances.* March 1964, 1.
108. Board of Water Commissioners, City and County of Denver. *The Blue River Diversion Project.* July 1964, 10.
109. *Features of the Denver Water System.* Office of Public Affairs, Denver Water Department, Denver, Colorado: December, 1976.
110. *Summit County Journal,* 21 April 1961, 1.
111. Board of Water Commissioners, City and County of Denver. *Final Construction Report for Dillon Dam and Appurtenances.* March 1964, 116.

www.ingramcontent.com/pod-product-compliance
Lightning Source LLC
Chambersburg PA
CBHW060030180426
43196CB00044B/2356